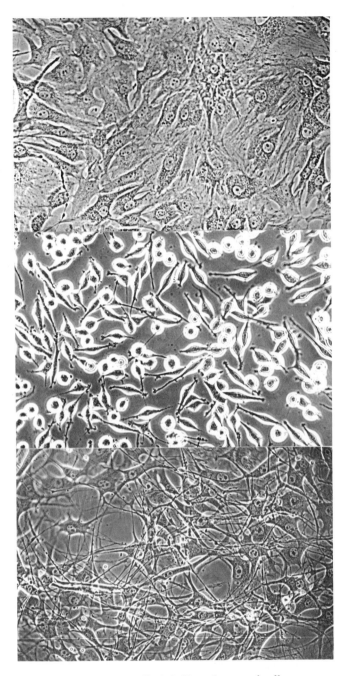

FRONTISPIECE: First interspecific hybrids and parental cells.
  Top: diploid rat cells; *middle*: mouse fibroblasts of line Cl.1D; *bottom*: hybrid cells MAT. From Weiss and Ephrussi,[256] with permission of *Genetics*.

HYBRIDIZATION OF
SOMATIC CELLS

BORIS EPHRUSSI

PRINCETON, NEW JERSEY
PRINCETON UNIVERSITY PRESS

575.28
Ep3h
82661
Man.1973

L.C. Card: 79-39783
I.S.B.N. 0-691-08114-x (hardcover edition)
I.S.B.N. 0-691-08117-4 (paperback edition)

This book has been composed in Linotype Baskerville

Printed in the United States of America
by Princeton University Press, Princeton, New Jersey

To

Brenda, Larry, Mary, Michi, *and* Rich

*who shared with me the joys of the
early days of cell hybridization,
when new perspectives were being opened to
studies of many fascinating biological phenomena.*

Preface

There may still be a use for people who believe there is more in biology than the applied biochemistry of the nucleic acids, always provided that they pay due regard to the man who has been trained to wield modern methods with precision and apply modern logical and mathematical facilities to the interpretation of his results.

(Sir Macfarlane Burnet. 1967, p. 2)

The body of this book is formed by the text of three Carter-Wallace Lectures on hybridization of somatic cells which I gave at Princeton University in January 1971. Their purpose and scope were briefly outlined at the beginning of the first lecture. At the risk of being somewhat redundant, I shall nevertheless define here the contents of the book more precisely.

Hybridization of somatic cells provides a recently developed technique and a methodologically new tool for the study of the genetic aspects of a variety of biological phenomena which can be observed in cells of higher animals *in vitro*. Recent years have witnessed its use in studies as varied as those of formal genetics of man, cytogenetics, regulation of macromolecular syntheses, embryology, virology, immunology, and cancer. As a result, by 1971, the literature on "cell hybrids" had grown to a point which made it both impossible and pointless to try to cover it in its entirety in three lectures. This, as well as the fact that the audience of the Carter-Wallace Lectures is very heterogeneous (with respect to background and degree of specialization), was taken into account in choosing a small number of topics of potential interest to the majority. This book is therefore neither a technical nor an exhaustive monograph on hybridization of somatic cells; it is rather an introduction to the principles of this new technique (presented in the course of an *historical account*), illustrated by some results obtained by its application in three areas of biological research, *formal genetics of man, cell differentiation,* and *cancer.*

The choice of these three areas, and the time allotted to the presentation of each, reflect a personal bias. For many years now, my interest has centered on the problem of the genetic mechan-

vii

isms involved in the differentiation of animal cells. I believe that somatic hybridization will be an important tool in the solution of this fundamental problem and that, in fact, a modest start in this direction has already been made. The presentation of the results obtained in this field therefore occupied nearly one-half of my lecture time.

Since I also believe that further progress in the understanding of cell differentiation will depend to a very large extent on the progress of formal genetics of higher animals, I thought it appropriate to begin by acquainting my audience with the way in which this branch of genetics is taking advantage of hybridization of somatic cells.

The end of my third lecture was devoted to a description of some new insights into cancer gained by hybridization of. neoplastic cells. I hope that it will become clear to the reader that this topic was chosen not because of the inevitable and justified interest generated by the words *cancer, malignancy*, and *neoplasia*, but because of the relevance of the mechanisms of tumor cell formation to the understanding of normal development and differentiation.

Needless to say, the bias of my presentation reflected also the desire to attract some of my young listeners to the type of research my co-workers and I have been conducting: hence a certain emphasis on our cellular rather than molecular approach to the study of cell differentiation.

A few additional words about the genesis of this book seem appropriate.

Invited in January 1969 to give the three Carter-Wallace Lectures, I gladly accepted to deliver them in May 1970; but circumstances prevented me from doing so. However, the Carter-Wallace Lectureship Committee generously renewed their invitation for the following year. I am grateful to the members of the Committee for the honor of these repeated invitations.

The preparation of the lectures obliged me to review, more carefully and critically than the pressures of everyday activity usually allow, the literature of my subject and, above all, to reevaluate the work which my co-workers and I have been and are pursuing. For unwittingly compelling me to do so, I am indebted to the Carter-Wallace Lectureship Committee.

To judge by the lively and fascinating discussions with students and faculty the week I spent at Princeton, my lectures were well received. This made my stay in Princeton a rewarding experience: it made me believe that there is truth in the words cited at the beginning of this Preface, and written by an outstanding biologist of my generation with whom I share the unrelenting passion to grasp the surely simple principles from which the astonishing complexity of living things emerges with surprising order. I have prepared these lectures for publication because of this belief.

My three Carter-Wallace Lectures, written before delivery, are reproduced below in almost unaltered form. I solicited this privilege both in the interest of unity of style and of rapid publication, without which the book would have been outdated before it was out of press. For granting me the privilege of this unusual procedure, I am grateful to the Princeton University Press.

While the text of my lectures thus appears below almost unchanged, a few additions were made during its preparation for publication. These comprise chiefly: some remarks which I did not have time to make during the lectures; some new data published or kindly communicated to me by their authors during the preparation of the manuscript for the press; and some references to work in areas which were not covered by my lectures. All of these additions were made to update the original text, to render it of *some* value to more specialized readers, and to indicate the relevance of some of the results to some more general biological problems. These additions appear either in small print and as footnotes in the text, or as numbered notes which will be found in the Appendix.* (The only notable exceptions are Chapters 7 and 8 which have been considerably expanded.) Lastly, I have added some figures (which were not shown as lantern slides during the lectures), and some epigraphs.

The better part of the work from my own laboratories in Gif (Laboratoire de Génétique Physiologique du CNRS, 1960–1961); in Cleveland, Ohio (Western Reserve University, 1962–1967); and

---

* I am aware of the inconvenience of Notes being placed at the end of the book. I have chosen this arrangement in order to present a coherent discussion of the main themes I wished to emphasize without digressions which are of interest only to more specialized readers.

at the Centre de Génétique Moléculaire du CNRS, 1967 to the present time, was done by my students and co-workers, all of whom I thank for their cooperation.

My thanks are also addressed to my many colleagues who allowed me to quote their as yet unpublished data, and to all those—authors, editors, and publishers—who granted me permissions to reproduce, and often generously supplied originals of their previously published figures, and/or to quote from their published writings. These contributions are acknowledged in the appropriate places.

The illustrations of publications from my present laboratory, as well as those prepared especially for my lectures and book, are the work of Mrs. S. Chevais whom I thank for her untiring co-operation.

The tedious job of obtaining all the required permissions and of typing the often illegible manuscript fell to the lot of Mrs. J. Lesmond, whose help is gratefully acknowledged.

Lastly, I wish to express a particular gratitude to Dr. M. C. Weiss for the many helpful discussions, the critical reading of the manuscript, and help in its preparation for the press.

Gif-sur-Yvette, September 1971

Contents

HYBRIDIZATION OF
SOMATIC CELLS

CHAPTER 1

Introduction

Except for Cassandra herself, prophets of doom have almost always turned out to be wrong, especially where scientific developments are concerned, and our remarks are not meant to discourage the great enthusiasm with which many molecular biologists are abandoning K12 for BALB C or some other mammal, such as a nematode.

(F. Jacob and J. Monod. 1970, p. 3)

In the initial letter of invitation addressed to me two years ago, the Chairman of the Carter-Wallace Lectureship Committee explained that the purpose of these lectures was "to call the attention of students and faculty in biochemistry, biology, and chemistry to exciting developments in important areas of the life sciences." I assumed that the development to which I owed the invitation was hybridization of somatic cells, and suggested this title for my lectures; and I take the renewed invitation* to lecture on this subject to mean that this theme had not come to be regarded meanwhile as less exciting, and hence, that the discouragement and pessimism which afflict nowadays so many biologists do not prevail at Princeton.

This pessimistic attitude was, as some of you may remember, signaled by Sol Spiegelman[240] in the late fifties, when he stated that "the outlook is depressingly bright for the quick resolution of many interesting problems" and, as you know, reached its apogee in Gunther Stent's most articulate book "The Coming of the Golden Age: a View of the End of Progress"[242] in which he claims that most of the essential principles of biology have been discovered and most of the essential problems solved; and that the few remaining ones are either not worth our bother or may not be solvable.

It appears to me true that, for the first time in the history of science, we have recently come to a point wherefrom one can see that science can eventually reach an end, but I do think that Stent's pessimism is premature, and I fully agree with the more

* See Preface.

constructive views recently expressed in two remarkable articles published in *Nature*: one entitled "Molecular Biology in the Year 2000" by Francis Crick,[37] the other "Genes and Hereditary Characteristics" by Alfred D. Hershey.[110] I highly recommend these two articles to those who are afraid that we shall soon run out of important problems, and I hope that they will feel somewhat reassured also by what I shall say in these lectures about the status of some fundamental biological questions.

Now, a few words about the plan of my lectures.

In the 10 years since its discovery in 1960, hybridization of somatic cells has grown from a biological curiosity into a method of analysis which is now so widely used in investigations of the genetic basis of a variety of biological phenomena, that an exhaustive review of the literature in a few lectures is no longer possible. I therefore think that it will be more rewarding if I limit my objectives: I shall first give you an *historical account* of the development of the method of somatic hybridization and of our knowledge of the properties of hybrid cells; and shall then speak in a somewhat more detailed way about three areas of biological research, presently investigated by means of somatic hybridization, which appear to me of particular importance or are of particular interest to me personally: *formal genetics, cell differentiation,* and *cancer* (Note 1).

CHAPTER 2

History of Somatic Hybridization

*I suspect that if we were honest we would have to admit
that if any one of us had never been, our science would
not have been quite the same; but it would be awfully
hard to see the difference.*

(Sir Macfarlane Burnet. 1967, p. 4)

I chose to start with an historical account of somatic cell hybridization for two reasons.

First, in my opinion, an honest historical account by an eye-witness inevitably shows the respective roles, in any important scientific development, of deliberate, logical design and of "lucky accidents" and, hence, especially in periods of pessimism about the future which many biologists are going through right now, inspires confidence and hope that the unexpected will, again and again, bring unforeseeable solutions to what, on logical grounds, appear today to be insoluble problems.

The second, and much less important reason, is that a history of somatic hybridization has been written recently by Henry Harris,[101] but his version is different from mine, and I may as well take this opportunity to present my own.

In presenting it to you, I shall assume that you are all acquainted, at least superficially, with the principles and techniques of cell culture *in vitro* on which hybridization of somatic cells relies, worked out chiefly by H. Eagle and T. T. Puck in this country. (The relevant references can be found in the excellent reviews by Green and Todaro[82] and by Krooth *et al.*,[144] and in the book of Morgan Harris;[108] a very brief account is given in Ref. 72. The definitions of some technical terms will be given in the Appendix.)

Since some of you may not be familiar with hybridization of somatic cells, I would like to begin by saying that the occurrence of this phenomenon was discovered in 1960 by a research team working in Paris under the leadership of Georges Barski,[5] and not by me as has been often erroneously stated in reviews of the

5

subject. I hasten to add that one person has not shared in this error: having made, in 1965, an important contribution to the field, Henry Harris, whose style sometimes makes one believe that he may be short on modesty, seems to ascribe the discovery of cell hybridization neither to Barski, nor to me, but to himself (Note 2).

The discovery of hybridization between cells of permanent lines

How was somatic hybridization discovered?

For several years, Barski had been studying two permanent, heteroploid lines of mouse cells growing *in vitro* and isolated many years earlier by Catherine Sanford and co-workers[214] at N.I.H. (Note 3). These two cell lines were derived from the same initial culture, i.e. were very closely related, but, in the course of years, had become very different in the degree of their neoplasticity. Having observed that the cells of the two lines differed both in their morphology and in their karyotype, Barski decided to look for evidence of Pneumococcus-like transformation by growing the two cell types together: what he apparently was looking for were morphological changes of cells whose identity could be established by their unchanged karyotype. What he discovered was quite different, however. It was the appearance, after three months of mixed culture, of a new cell type which was characterized *quantitatively* by a total number of chromosomes nearly equal to the sum of the modal chromosome numbers of each of the parental lines, and *qualitatively* by the presence of essentially all of the marker chromosomes of the two parents, i.e. of the chromosomes which differentiated one cell type from the other. These were, clearly, hybrid cells: they contained in a single nucleus the combined genomes of the two parents.[5, 6]

For a long time, karyological analysis was the only means of identifying hybrid cells, and, as you will see, the correlation of phenotypic traits with the karyotype is one of the most important aspects of somatic cell genetics; therefore, I think that it is worth showing you three microphotographs which illustrate Barski's first experiments (Note 4).

Figure 2.1 shows the karyotypes of the parental cells of this "cross" and of the resulting hybrids. In 2.1A is shown a metaphase of a cell of (the highly malignant) line N-1, with 54 acrocentric

6

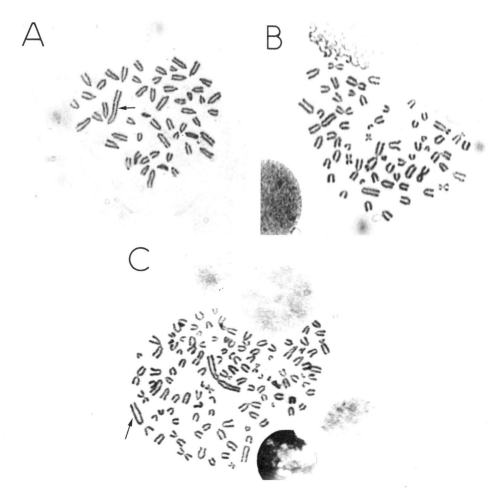

FIGURE 2.1. Metaphases of parental and hybrid cells of Barski's first cross. *A*, N–1; *B*, N–2; *C*, hybrid. From Barski *et al.*[6] Courtesy of G. Barski.

chromosomes, including an extra long one. Figure 2.1B shows a metaphase of a cell of the other "parent" line, N-2 (of low malignancy): it contains 61 chromosomes, 12 of which are biarmed (or metacentric). Figure 2.1C shows a metaphase of a hybrid cell with a total number of 113 chromosomes; among them you will recognize the marker chromosomes of this cross: the extra long acrocentric chromosome of N-1 and the 12 biarmed chromosomes of N-2.

7

Barksi's first publications[5, 6] of these results were received with great scepticism. Because of their apparent rarity, hybrid metaphases, observed *alongside* metaphases of the parental cells, were regarded by most karyologists as artifacts resulting from the superposition of two parental metaphases in the karyological preparations. However, it appeared to me that, if true, Barski's findings were of sufficient potential importance to deserve verification. This I set out to do with my co-workers in early 1960, using two close relatives of the cell lines used by Barski, and within a few months we reproduced Barski's results.[238]

Generality of the phenomenon

Having convinced myself of the reality of the phenomenon, having furthermore isolated pure hybrid populations from this cross (this was a relatively easy task since the hybrids had a selective advantage over the parental cells), and having observed some of the properties of the hybrid cells—I set out with great optimism to prove that somatic hybridization can be forged into a tool of genetic analysis of somatic cells.

What had to be shown to that end, as I saw it then, could be stated in three points.[66] It had to be proved that:

1. It is possible to obtain hybrids between any two cells differing in some or many of their genetic properties.

2. Such hybrid cells can be isolated or preferably selected and maintained in pure culture so as to make the study of their phenotypes possible.

3. The formation of hybrid cells is followed by some process—be it only mitotic abnormalities resulting in the accidental loss of chromosomes—which leads to the formation of cells with different constellations of genes, that is (using the term very loosely), in gene or chromosome segregation.

I must admit that my optimism was based on the rather shaky ground of our first observations of the few somatic hybrids obtained. By now, however, all of these conditions have, to a great extent, been fulfilled, and I will now describe the different steps whereby somatic hybridization reached its present status.

Concerning the first point (i.e. the possibility of obtaining hybrids between genetically different cells), it became clear within the next couple of years that the "spontaneous" formation of

8

hybrid cells, containing in a single nucleus the quasi-complete genomes of both parents, occurs, as a rare event, in mixed cultures of practically any two types of karyologically distinguishable heteroploid cells of the so-called "permanent lines," and that these hybrids can be isolated, either owing to their selective advantage or by cloning.[61, 69, 77] This was first shown by crosses between cells of several pairs of permanent mouse lines of different origins performed in my laboratory (review in Ref. 61). Thus, by 1963, we had growing as pure cultures a number of karyologically identified hybrids between genetically different cells: I say "genetically different" because the cell lines involved had been derived from inbred strains of mice differing in histocompatibility genes and, surely, many other genes. This, in turn, permitted the first studies of the expression, in the hybrids, of the genes by which the parental cells differed. It was thus shown, to begin with, that as expected, these hybrids exhibit the genetic traits of both parents. For example, hybrids produced by the cross of cells having two different H-2 histocompatibility antigen complexes, carry, on their surfaces, the antigens characteristic of both parents: there is co-dominance of the genes specifying these antigens.[77, 239] Similarly, by crossing two mouse cell lines, derived from mice in which the enzyme β-glucuronidase differs in heat stability, hybrids were obtained which synthesized both "allelic" forms of β-glucuronidase.[76] These were the first proofs that both parental genomes are active in somatic hybrids, and numerous similar observations have been made in subsequent years.

Crosses involving normal diploid cells

What I said about the quasi-general occurrence of hybridization between cells of permanent cell lines should not make you think that every cross we attempted resulted in success. We did have a certain number of failures. Possibly they were due to the rarity of hybrids between some cell types or, more probably, to the lack of selective advantage of certain hybrids over the parental cells. Indeed, we learned later by pure accident that many young hybrids, for some unknown reason, are incapable, to begin with, of overgrowing the parental cells at 37°, while they do so at 29° Several new hybrids were isolated in this way,[61, 267] including one of a new type which was of particular importance. It was a first hybrid[217] between cells of a permanent line of mouse cells (2472-6, derived from N-1, already familiar to you) and *diploid* cells from a newborn mouse carrying a small chromosome translocation (T-6). The karyotypes of the parents and of the hybrid are shown in Figure 2.2. The metaphase shown in A is that of a cell of line

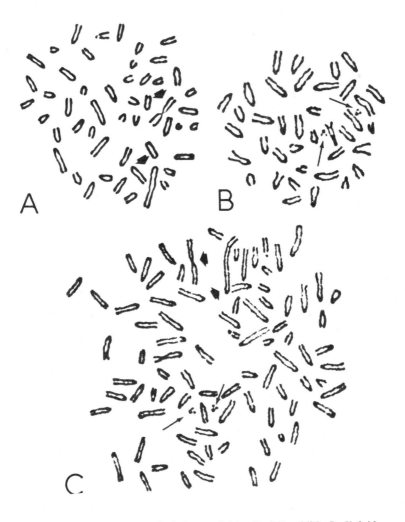

FIGURE 2.2. Metaphases of: *A*, heteroploid cell of line 2472; *B*, diploid cell carrying the T–6 translocation; *C*, hybrid cell. Arrows indicate marker chromosomes. Reproduced from Ephrussi.[61] Copyright 1966, The Williams and Wilkins Company, Baltimore, Maryland, with permission.

10

2472-6. It comprises 50 chromosomes among which you will recognize the extra long acrocentric chromosome of N-1. You will notice that it also contains now a long metacentric chromosome which provides another marker in the cross with the diploid mouse cells. The karyotype of the latter, shown in B, comprises the normal complement of 40 chromosomes, all clearly acrocentric except for the two very characteristic "translocated" (T-6) chromosomes. The marker chromosomes of both parents are easily recognized in the metaphase of the hybrid cell, shown in C. The total chromosome number here is 89, which is very nearly the expected one.

This hybrid was of particular importance because it showed, for the first time, that diploid cells can be hybridized also, at least when their partners are cells of a permanent, neoplastic line. Moreover, similar crosses performed not with freshly explanted T-6 diploid cells, but with senescent cells of the same origin, hardly dividing any more, gave rise to similar hybrids. In both cases, they were shown to inherit from the permanent parent the *lack of contact inhibition* (or, in other words, the ability of multi-layered growth, characteristic of permanent lines) as well as the *capacity of rapid and continuous growth*.[45, 217] The question of the malignancy of such hybrids will be briefly discussed in my last lecture.

A first selective system for the isolation of somatic hybrids

Returning to the selection of hybrids at low temperature: in view of several successes of this method, I at one time had high hopes that we had "put our finger" on a universal system for the selection of hybrids. These hopes did not materialize, however. But the mention of this fact brings me to the next important development—the demonstration by John Littlefield[150, 151] in 1964 that by using two cell lines, carrying different selective markers, one can, as in microbial genetics, establish a selective system wherein the hybrid cells will grow and the parental cells will not. Littlefield's system is, in fact, an adaptation to cell hybridization of a selective system designed by Szybalski and co-workers for other purposes.[243]

Littlefield selected two sublines of the heteroploid mouse line known as the L-line: one resistant to 8-azaguanine, the other to

11

5-bromodeoxyuridine (BUdR).* These two drug resistances are correlated, respectively, with deficiencies for the enzymes hypoxanthine-guanine phosphorybosyl transferase (HGPRT) and thymidine kinase (TK) which are required for phosphorylation (and therefore for incorporation) of the base analogs. The system of hybrid selection is based on the fact that mammalian cells have open to them two pathways of synthesis of nucleotides (Figure 2.3): the *de novo* pathway whereby nucleotides are synthesized from sugars and amino acids, and the "scavenger" pathway which utilizes the preformed nucleosides, hypoxanthine and thymidine. The *de novo* pathway can be blocked by aminopterin. The operation of the scavenger pathway depends on the simultaneous presence of TK and HGPRT. Therefore, the drug

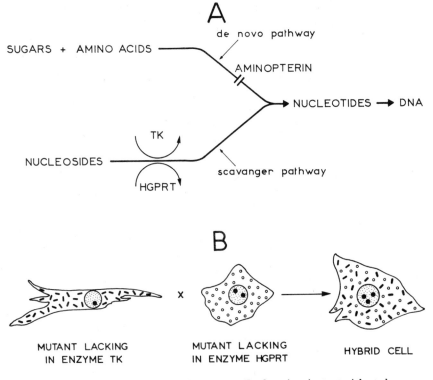

FIGURE 2.3. Littlefield's selective system. Explanation in text. Adapted from "Hybrid Somatic Cells" by Boris Ephrussi and Mary C. Weiss. Copyright © 1969 by *Scientific American, Inc.* All rights reserved.

* 3 μg/ml and 30 μg/ml, respectively. Most of the other cell lines, said to be resistant to 8-azaguanine and BUdR, present similar resistance levels.

resistant (enzyme deficient) parental cells are unable to grow in a medium (HAT) containing hypoxanthine, aminopterin, and thymidine. On the contrary, the hybrids which contain the genes of both parents, and therefore produce both TK and HGPRT, grow unhampered in HAT (Figure 2.3B).

Notice that this experiment provided the first demonstration of the occurrence of complementation in hybrids of mammalian cells, i.e. the continued production in the hybrid cells of the two enzymes, each of which is synthesized by only one of the parents.* As expected in this situation, we have here dominance, at the phenotypic level, of the normal condition over the enzymatic deficiency.

Notice also that Littlefield's system permits the quantification of hybrid formation. The spontaneous "mating rate" determined in his cross between L cells was of the order of 10^{-5}. We shall soon see, however, on the one hand, that much higher mating rates can now be obtained, and on the other, that the exact significance of the term "mating rate" requires qualification.

Meanwhile, I would like to emphasize the importance of Littlefield's contribution. Very soon his system was used as a model in my own and several other laboratories: similar drug resistance markers have been introduced into several other permanent cell lines and hybrids obtained by analogous procedures.

A semi-selective system

The next important technical development was the establishment in 1965, by Davidson and me,[45] of a modification of Littlefield's system, a modification which became known as the "half-selective" system. We showed that, by the use of the same selective medium, HAT, one can easily obtain hybrids between heteroploid cells, carrying a selective marker, with *unmarked* diploid cells. This is effective because the heteroploid cells are killed by the selective medium, while the hybrids possess the enzymes necessary for

---

* The cell lines used by Littlefield were sublines of the same L-line and unfortunately presented no reliable karyological differences. Thus karyology could provide no independent proof of the hybrid nature of the selected (2s) clones. Such a confirmation was, however, provided by Mary Weiss who reproduced, in my laboratory, Littlefield's experiments using a different line of thymidine kinase deficient L cells (unpublished results, quoted in Ref. 63).

growth in HAT and give rise to thick colonies which usually can be easily distinguished from the very slow growing and thin colonies formed by the diploid (parent) cells. Figure 2.4 shows how the system looks in reality: it is a photograph of a culture bottle from our very first attempt at isolating hybrids by the use of the half-selective system. Many of the thick (darkly stained) colonies which can be seen growing on its bottom were isolated, subcultured, and analyzed karyologically: all of them proved to be colonies of hybrid cells.

Experiments of the same type have shown that if one uses *very unequal proportions* of parental cells (say, 10^6 cells of the "marked" line and only 10^2 or 10^3 diploid cells), the "effective mating rate" is for some reason very high: 1 out of 100 diploid cells forms a hybrid.[46] As the proportion of diploid cells is increased, the "effective mating rate"* of these cells decreases. (This unexplained effect is also observed in other spontaneous crosses and in virus induced hybridization to be described below.)

The "half selective system" is obviously of a much wider applicability than Littlefield's original system: no wonder it has been and is being widely used, as we shall see.

Summing up at this point, I may say that by the end of 1964, it had been convincingly shown that somatic hybridization can be detected between most pairs of cells of karyologically different permanent mouse lines; that diploid mouse cells can be crossed with cells of permanent lines; that such hybrids, isolated in pure culture, are capable of indefinite multiplication; that genetic traits of both parents are expressed in the hybrids; that in these hybrids dominance, co-dominance, recessiveness, and complementation of genetic markers are observed as expected; and finally, that the principles of a selective system, and of a semi-selective system, both allowing quantification of the phenomenon, had been established. I may add that the karyotypic evolution of hybrid cells had been studied, but the resulting observations will be more conveniently discussed later (pp. 28f.); and that most of the above data had been published by the end of 1964.

* This term was introduced to indicate that the observed number of hybrid colonies is a measure of the frequency of formation of *viable* fusion products and not necessarily of the frequency of cell fusion.

14

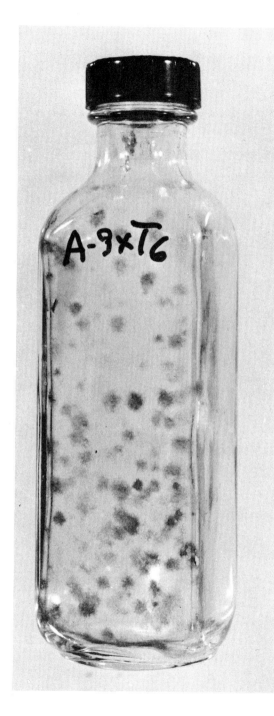

FIGURE 2.4. Culture bottle showing hybrid colonies obtained by the use of the semi-selective system.[45]

Virus induced cell fusion and heterokaryon formation

It is at this stage and with this background that a new and important development occurred: the introduction of virus-induced cell fusion, which was to greatly facilitate and to extend the range of somatic hybridization. I was at that time working in Cleveland; glancing one morning through my newspaper, I learned of Henry Harris' spectacular entrance on the scene. An attractive advertisement in the *New York Times* of February 17, 1965 caught my attention. Shifting my eye to the two-column article printed next to it, I discovered that it was a rather sensational account of a paper published by Harris and Watkins a few days earlier in *Nature*, under the title: "Hybrid cells derived from mouse and man: artificial heterokaryons of mammalian cells from different species."[106] In essence, the original paper, which I read a few days later, announced without any reference whatsoever to any of the published groundwork I have described, that treatment of mixed suspensions of human (HeLa) and mouse (Ehrlich ascites) cells, with UV inactivated hemagglutinating virus of Japan (HVJ, now currently known as Sendaï virus),* earlier shown by Okada[184] and Okada and Tadakoro[187] to produce fusion of Ehrlich ascites cells, results in the formation of numerous homo- and heterokaryons. The latter contain, in a single mass of cytoplasm, various numbers of nuclei of the two cell types. Harris and Watkins also showed that the heterokaryons continue for some days to metabolize and sometimes undergo nuclear fusion, followed, in very rare instances, by mostly abortive mitoses.

Shortly thereafter, Harris showed that differentiated cells, whether of the same or of different species, can also be fused by the virus to form metabolizing heterokaryons.[98]

Let me illustrate what I have told you with a few microphotographs.

The photographs of Figure 2.5 show virus induced agglutination of Ehrlich ascites cells in suspension. These photographs are taken from a paper published in 1962 by Okada,[185] from whom Harris and Watkins borrowed the technique. The latter consists in adding to a cell suspension some live or UV inactivated Sendaï virus, and shaking the mixture for a short time in the cold. As

---

* An RNA-containing virus.

16

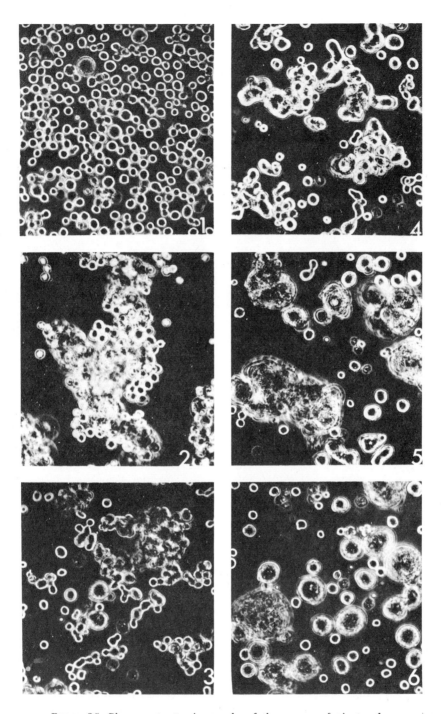

FIGURE 2.5. Phase contrast micrographs of the process of giant polynuclear cell formation caused by HVJ from Ehrlich ascites tumor
cells. *1*, original Ehrlich tumor cells (control); *2*, cell agglutination by
addition of HVJ at 4°C; *3*, after 5 min incubation at 37°C; cell aggregates show the first step of cell fusion; *4*, 10 min after the onset of 37°C
incubation; *5*, 20 min after the onset of 37°C incubation; *6*, 30 min after
the onset of 37°C incubation. Typical fused giant cells are shown. From
Okada.[185] Copyright 1962, Academic Press, New York, with permission.

you can see in Figure 2.5, the agglutinating action of the virus causes the formation of clumps of cells of various sizes. When the cells are placed at 37°, this agglutination is soon followed by fusion of some of the cells, a process which is shown in Figures 2.6 and

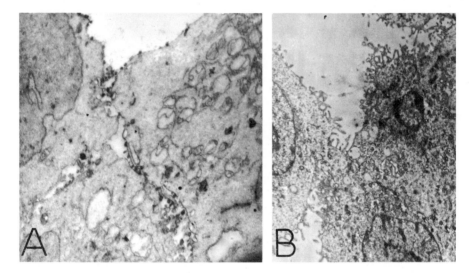

FIGURE 2.6. Electron micrographs showing two successive stages of Sendaï virus induced fusion of Ehrlich ascites cells. *A*, 20 min after addition of virus at 4°C: many virus particles adsorbed onto the surfaces of agglutinated cells are visible; *B*, after 5 min incubation at 37°C: connections between two cells, covered with complex surface projections, can be seen. From Okada.[185] Copyright 1962, Academic Press, New York, with permission.

2.7. Figures 2.6A and B are electron micrographs of two fusing Ehrlich ascites cells (also published by Okada in 1962[185]): one can see the formation, under the action of the virus, of intercellular bridges. These soon cause the confluence of the cytoplasms of two or more cells: thus a single mass of cytoplasm is formed containing practically any number of nuclei. An example of this is shown in Figure 2.7 which is taken from a paper published by Okada in 1958.[184]

Okada, interested chiefly in the mechanism of virus-induced cell fusion,[186] did not push his investigations further. Harris and his group were, on the contrary, interested in the fate of the fused cells. Therefore, they fused a variety of cells of different types and species with the help of UV inactivated Sendaï virus. They then

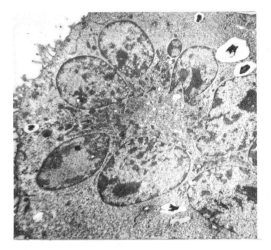

FIGURE 2.7. Electron micrograph of a polykaryocyte resulting from virus induced fusion of Ehrlich ascites cells. From Okada[184] with permission of *Biken's Journal*.

let them attach to coverslips (placed in Petri dishes containing nutrient medium) on which the cells spread out within the next 24 hours, permitting observations on their properties and evolution in the course of the following days. Figures 2.8 and 2.9, taken from the first extensive account of the observations of Harris and co-workers,[107] show some examples of heterokaryons formed by the fusion of cells with nuclei which are either morphologically different or are identified by labelling with tritiated thymidine prior to fusion.

Figure 2.8 shows heterokaryons resulting from the fusion of Ehrlich ascites (mouse) cells with cells of the well-known (human) permanent line HeLa. In the heterokaryon shown in 2.8A, the two upper nuclei, each with two large nucleoli, are derived from HeLa cells; the two lower, and darker nuclei, from Ehrlich ascites cells. The heterokaryon shown in Figure 2.8B contains three prelabeled HeLa nuclei, and one unlabeled Ehrlich ascites nucleus.

Figure 2.9A shows a heterokaryon formed by fusion of one HeLa cell with nine rabbit macrophages. The large nucleus is a HeLa nucleus, the small ones are macrophage nuclei.

The heterokaryon shown in Figure 2.9B contains four rabbit macrophage nuclei and three much smaller hen erythrocyte nuclei.

19

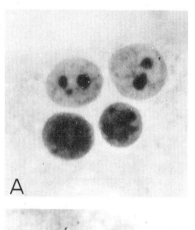

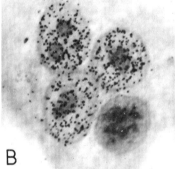

FIGURE 2.8. *A*, Tetranucleate cell in which the two upper nuclei are derived from HeLa cells and the two lower ones from Ehrlich ascites cells; *B*, autoradiograph of a tetranucleate cell containing three HeLa nuclei, labelled with tritiated thymidine prior to fusion, and one unlabelled Ehrlich ascites nucleus.[107] Courtesy of Henry Harris.

20

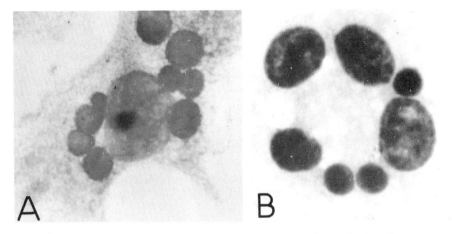

FIGURE 2.9. *A*, Heterokaryon containing one HeLa nucleus and a number of rabbit macrophage nuclei; *B*, heterokaryon containing four rabbit macrophage nuclei and three much smaller hen erythrocyte nuclei.[107] Courtesy of Henry Harris.

Although these heterokaryons were qualified as "viable" (Note 5), they have a very limited life span, and are therefore of rather limited use (Note 6). They were, however, the source of some interesting observations by Harris and his co-workers, about which I am going to talk in my next lecture.

Concerning the production, by the use of Sendaï virus, of cell hybrids capable of multiplication, necessary for genetic analysis, I shall say at this point only that it was not immediately achieved. Harris' lack of success in this respect was due, I think, chiefly to an unfortunate choice of cell lines: Ehrlich ascites and HeLa cells. This combination has produced no viable hybrids to this day for reasons which were rather clear from Harris and Watkins' very first observations[106] which showed that the Ehrlich nuclei rapidly and almost totally suppress DNA synthesis in HeLa nuclei (a fact later confirmed by a detailed study by Johnson and Harris[124]).

Nevertheless, Harris and Watkins' early work had two important effects on the progress of cell hybridization. It provided:

(1) a stimulus for the isolation of the first spontaneous, *interspecific* hybrids capable of indefinite multiplication; and

(2) a stimulus to perfect the Sendaï technique by combining it with previously established procedures (see above). We shall see that only thereafter did virus induced fusion come to full fruition and result in a number of important observations made in Harris' own laboratory, as well as in many others, and thus prove to be of undeniable value.

Production of viable hybrids by the use of Sendaï virus

Let me first say a few words on the second point.

The possibility of producing viable hybrids, either intra- or interspecific, by the use of Sendaï virus, was soon suggested by experiments of Yerganian and Nell[266] and definitely proved by Coon and Weiss,[35, 36] who, using both of the quantitative selective techniques I described earlier, made a systematic study of the optimal conditions for use of the virus conducive to the production of a maximal yield of viable hybrids. The upshot of these studies was that under favorably chosen conditions, the yield of viable hybrids can be increased about a hundredfold as compared

22

with spontaneous hybridization. If proper ratios of parental cells characterized by different fusion capacities and, above all, proper virus concentrations are chosen (so as to result in the formation of the maximum number of di-heterokaryons), it becomes possible in favorable cases, to use virus-induced fusion as an *alternative to selection* (Note 7). It goes without saying, however, that the presence of selective markers greatly facilitates the isolation of hybrids. I may add that Coon and Weiss have shown also that the *surviving* hybrids do not differ from spontaneous hybrids either in their karyological properties, or in their ability of continuous multiplication.* I say "surviving hybrids" because it is only a fraction (probably of the order of 10%) of all di-heterokaryons which give rise to mononucleate hybrids capable of multiplication (Note 8).

Interspecific somatic hybrids

Returning now to the first point (page 22), let me first repeat that it was Harris and Watkins' demonstration that cells of different species can be fused with the help of Sendaï virus that in 1965 led Mary Weiss and me to the isolation of the first viable interspecific hybrids.[70, 256] What we did was simply to apply the half-selective system to mixed cultures of mouse fibroblasts, carrying a selective marker, and of freshly explanted, unmarked diploid rat cells. The isolation of spontaneous hybrids proved to be amazingly easy. I shall show you pictures of these hybrids in a minute; but before I do so I would like to emphasize that we undertook interspecific crosses not as a matter of curiosity, but in a deliberate attempt to bypass two defects of intraspecific crosses, namely:

1. the similarity, if not identity, of the chromosomes of the parents of intraspecific hybrids;

2. the scarcity of genetic markers distinguishing animals (and, hence, cells) of the same species.

We knew that, if interspecific crosses proved to be feasible, we could choose two species karyologically so different that most (and sometimes all) chromosomes in the hybrid cells could be

---

* It is still uncertain whether the so-called spontaneous hybridization is due to virus (es) present in the parental cells.

identified as to species of origin.* We also knew that many homologous enzymes of different animal species can be identified as to species of origin by simple physico-chemical procedures such as electrophoresis.[161] And we therefore hoped that, if interspecific hybrids proved to be viable, and if both parental genomes remained active in these hybrids, they would contain very numerous genetic markers in the form of homologous proteins recognizable by their physico-chemical properties. These hopes proved to be justified.

Let me illustrate these statements with some pictures. The top photograph of the frontispiece shows a thin monolayer of flat, contact inhibited diploid rat cells (R_1); the middle one, the refractile and more loosely attached cells of the heteroploid, thymidine kinase deficient mouse line LM (TK$^-$) Clone 1D (which I shall call, from now on, simply Clone 1D or Cl.1D); the lower photograph shows the obviously happily growing mouse $\times$ rat hybrids which we called MAT and which form a loose multi-layer of cells whose morphology and growth habit are intermediate between those of the parental cells.

As will be seen in Figure 2.10, karyological analysis of these three cell types provided unmistakable proof of the hybrid nature of MAT. In this figure, the karyotypes of the cells are shown in two forms: as metaphases on the left and as karyograms on the right. (In karyograms, the chromosomes of the metaphase are grouped according to size and shape.) You will notice that: (1) the parental cells differ in the number and shape of chromosomes; (2) some of the chromosomes of each parent (indicated by arrows) represent particularly good karyological markers; (3) the hybrid cells contain all marker chromosomes of the parents. The total number of chromosomes in the hybrid shown in this figure is 89, which is just below the expected range: the diploid rat cells have 42 chromosomes; the *modal* chromosome number in Cl.1D is 53 (range: 51–54).

The first demonstration that both parental genomes are active in interspecific hybrids was provided by the study of the electrophoretic patterns (on cellulose acetate strips) of lactate dehydro-

---

* Barring, of course, chromosome rearrangements which proved to be relatively rare. Some notable exceptions will be dealt with at the end of the next chapter.

24

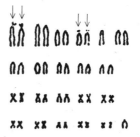

FIGURE 2.10. Karyotypes of rat, mouse, and hybrid cells. On the *left*, metaphases, on the *right*, karyograms. The two top pictures show the karyotype of the rat cells, the two middle ones that of the mouse cells, and the two bottom ones that of hybrid cells MAT. Reproduced from Ephrussi and Weiss[72] with permission of *Scientific American*. The lower right hand photograph is reproduced with the permission of *Genetics* where it had been previously published.[256]

genase (LDH) from the parental (mouse and rat) and hybrid cells.[257] It will be recalled that LDH is a tetramer formed by the random association of two different types of subunits, A and B, specified by two nonallelic genes. This association results in the formation of five different combinations (B_4, B_3A_1, B_2A_2, B_1A_3, and A_4) or isozymes, numbered 1 to 5.

The electrophoretic mobilities of rat and mouse LDH isozymes are different: this is clearly shown in Figure 2.11 for LDH ex-

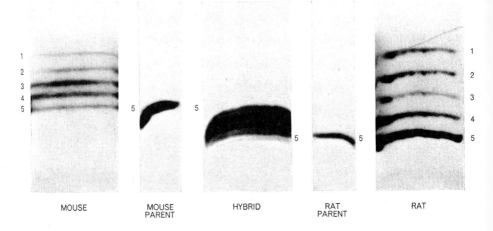

MOUSE   MOUSE PARENT   HYBRID   RAT PARENT   RAT

FIGURE 2.11. Electrophoretograms of lactate dehydrogenase of diaphragm and of *in vitro* cultured cells. The gels on the extreme *left* and extreme *right*, are of extracts of mouse and rat diaphragm, respectively. No. 5 bands are the only ones present in the parental cells, and differ in electrophoretic mobility. The hybrids have both parental bands and also three intermediate bands, representing hybrid molecules (*center*). From Ephrussi and Weiss[72] with permission of *Scientific American*.

tracted from rat and mouse diaphragms (this tissue was chosen because, unlike many others, it displays all five LDH isozymes). It will be noticed that the difference in electrophoretic mobility is especially clear in the case of LDH-5, formed by the association of four identical subunits of type A.

Cells grown *in vitro* very often contain only LDH-5. This is true of the parental cells with which we are concerned here: on the zymograms, each displays a single clear band corresponding to LDH-5 of the parental species (Figure 2.11). The LDH pattern of the hybrids clearly shows that these cells produce the LDH-5 bands characteristic of both parents and, in addition, new bands of "hybrid enzymes" resulting from the association of parental (type A) subunits. I must emphasize that hybrid bands do *not* appear in mixtures of extracts of parental cells.

The examination of β-glucuronidase in the parental mouse and rat cells, and in their hybrids also showed the activity, in the hybrids, of genes of both species (these enzymes were identified both by different electrophoretic mobilities and thermostabilities).[257]

I shall add here two remarks, the first of technical, the second of theoretical interest.

First, similar evidence for the activity of both parental genomes in a number of different interspecific hybrids has by now been provided by the study of a number of other enzyme proteins which therefore represent most convenient "inbuilt genetic markers." This explains why interspecific hybrids, as you will see, have come to play a particularly important role in the study of somatic cell genetics.

Second, the significant conclusion drawn from our first observations on rat × mouse hybrids was that between somatic cells of these species, there is no incompatibility similar to that observed in sexual interspecific crosses,[256] and that the usual lethality of sexual hybrids between animal species when they do contain both parental genomes must be due to the inadequacy of their nuclei for directing normal development and differentiation, rather than to their inability to coordinate the various molecular events of the cell cycle.[62, 71] The maintenance of the quasi-total and functional chromosomal complements of the two species in the course of prolonged multiplication of the first interspecific hybrids[256] showed, on the contrary, that the signals which govern the different phases of the cell cycle are devoid of species specificity and thus assure, in particular, the coordinated replication of the chromosomes of the two combined species. This conclusion remains valid for a number of other hybrids between species, but we shall see also that it breaks down in certain cases; and that it is both the rule and the exceptions which have made interspecific hybrids the most favored material for the study of a number of problems.

27

Human × mouse hybrids

These remarks bring me to the last major "breakthrough" in cell hybridization which resulted from the production of the first viable human × mouse hybrids and from their karyological study.

As I said earlier, when I went into cell hybridization I had hoped that the formation of highly polyploid hybrid cells would be followed by chromosome losses, if only due to frequent mitotic irregularities, which might be fatal to the genetically balanced parental cells, but would be compatible with the survival of segregants of somatic hybrids because they must, to begin with, contain a number of genes in excess of the required minimum.

In fact, all the hybrids I have talked about thus far lose *some* chromosomes during their growth *in vitro*. In *intra*specific (mouse × mouse) hybrids, this loss (apparently a random process, probably due to mitotic errors) is slow, and is not extensive. After a couple of hundreds of cell generations, mouse × mouse hybrids usually lose 10–20% of the total chromosome complement —then the hybrid karyotype appears to be stabilized.[60, 66, 68, 69]

I must mention, however, that according to a recent report by Engel *et al.*,[58, 59] more important losses occur in mouse × mouse hybrids with much more prolonged growth, at least under the conditions of cultivation used by these authors (Note 9).

In most *inter*specific hybrids between not too remote species (such as mouse and rat, or mouse and Chinese hamster), there is a *slight preferential* loss of choromosomes of one of the species: rat chromosomes in rat × mouse hybrids;[256] mouse chromosomes in hamster × mouse hybrids.[218] However, the extent of the total loss here also is limited and of the order of 10–20%.

From the point of view of genetic analysis, these results were not too encouraging; hence, a few attempts at inducing or accelerating the slow, spontaneous loss of chromosomes in mouse × mouse hybrids were made in my laboratory. They were unsuccessful, however. In 1967, Mary Weiss, then working on collagen synthesis in Howard Green's laboratory at NYU, isolated the first spontaneous, viable human × mouse hybrids and, to her surprise, discovered that they were subject to extremely rapid, preferential loss of human chromosomes[259]—so rapid, in fact, that the initial hybrid, with a full complement of both mouse and human chromosomes, was not observed until much later. In the particular cross studied by Weiss and Green,[259] between Clone 1D cells and diploid human cells, the earliest hybrid cells observed (*ca.* 20 generations old) contained the full complement of mouse chromosomes and only 2 to 15 human chromosomes; this initial rapid loss was followed by a slower decline; then, after 100 to 150 generations, stabilization at some 1 to 3 human chromosomes.

As you will see, a variety of other human $\times$ mouse crosses have been performed since: in all cases, there was found to be a preferential loss of human chromosomes, but the rate of loss appears to be somewhat different in different hybrids.

A similar loss of human chromosomes from hybrids between Chinese hamster cells and human cells has recently been observed by Kao and Puck.[132]

Although we still know nothing about the mechanism of this remarkable phenomenon, two things are clear: (1) it undoubtedly represents a case of breakdown of the mechanisms which in other hybrids insure the co-ordinated transmission of the two parental genomes to their daughter cells; it therefore offers excellent opportunities for the study of the mechanism of this co-ordination[71] (Note 10); (2) the discovery of this phenomenon has been the starting point of a number of developments, in particular of the study of formal genetics of man, by means of somatic hybridization, about which I shall speak in the next lecture (Chapter 3).

Present status of somatic hybridization

I have given you the principal stages of the development of somatic hybridization. It only remains to bring you up on its present status. This can be done in a few words. The following have been achieved by now:

1. Several new selective systems have been established.

2. Inborn errors of metabolism[144] have been used to produce hybrids between human cells.

3. Rapid chromosome loss has been observed to occur in certain other interspecific hybrids.

4. Recent unpublished work by Pontecorvo has shown that

loss of the chromosomes of one of the parents can be experimentally induced.*

The combined use of these facts and techniques permits today
the production of practically any hybrid one wishes to have for
any purpose.

And the upshot of it is that, like real molecular biologists, we
now receive with fear each new issue of *P.N.A.S.*!

* For some details on these four points and relevant references, see Appendix, Notes 11–14.

30

CHAPTER 3

Application of Cell Hybridization
to the Study of Formal Genetics

*. . . it is clear that, no matter what the difficulties and
the disappointments, if we want a breakthrough in human
genetics we have to concentrate on methods which bypass
sexual reproduction.*

(G. Pontecorvo. 1961, p. 20)

*There is, of course, a large field where at present we can
only speak in terms of probability. This is merely another
way of saying that we are ignorant.*

(J.B.S. Haldane. 1952, p. 25)

Personally, I have little taste for formal genetics in general, let
alone that of man. But formal genetics is beyond any possible
doubt an absolutely essential prerequisite to the attack on any
other problem involving genetic mechanisms and that is why I
must give you at least an idea of the principles on which formal
genetic analysis of somatic cells is based. Since it is formal genetics
of man that is at present making the most rapid strides, I shall
speak exclusively about that. I wish to emphasize, however, that
with the introduction of some new techniques to be mentioned
later (p. 127), the principles which I shall describe should become
applicable to the study of formal genetics of other mammals as
well.

Formal genetics of our own, among all mammalian species, is
now making the fastest progress not simply because of more fund-
ing, but because extremely rapid and preferential loss of chromosomes of one species from interspecific hybrids was first discovered
in human $\times$ mouse hybrids; and the chromosomes lost happened
to be the human ones. While in all other interspecific hybrids
thus far mentioned, the persistence of two functionally active
genomes permits us, in principle, to establish allelism, dominance,
co-dominance, recessiveness, complementation; the rapid loss of
most of the human chromosomes in human $\times$ mouse hybrids
enables us, in addition, to correlate the presence of specific human

31

gene products with the presence of definite human chromosomes. Moreover, granted a sufficient number of markers (and we know that interspecific hybrids always carry a large number of "inbuilt markers"), it is possible to establish human linkage groups and to assign them to definite chromosomes.

The first localization of a human gene by means of somatic hybridization

Let me go back to the first cross between mouse and human cells performed by Weiss and Green,[259] and already briefly mentioned. This cross involved, on the one hand, the (BUdR resistant, thymidine kinase deficient) mouse line Cl.1D, already known to you, and a line of normal human diploid cells (WI-38). The cross was initiated by setting up a mixed culture, the mouse cells being in great excess (2×10^6 mouse cells as against 10^4 human cells). The spontaneously formed hybrids were isolated by the use of the "half-selective system," i.e. in HAT medium. You remember that this was possible because the thymidine kinase deficient mouse cells are killed in HAT; because the slow growing human diploid cells, although resistant to HAT, form only a thin monolayer; and because hybrid cells contain the human $TK+$ gene introduced by the human parent: they therefore do grow in the selective medium wherein they form discrete, multilayered colonies. However, these hybrids have morphological characteristics of both parents as can be seen on the left side of Figure 3.1 which shows how the cultures of the three types of cells look.

The karyotypes of the three types of cells are shown on the right hand side of Figure 3.1. The uppermost karyogram is of the mouse line Cl.1D. There are 52 chromosomes in all. Nine of them are long bi-armed chromosomes; among these you will notice one (arrow) which is easily recognizable owing to a secondary constriction ("chromosome D"). The remaining chromosomes are acrocentric, like all those of normal mouse cells.

The middle karyogram in Figure 3.1 is of a normal diploid (female) human cell with 46 chromosomes, arranged in 7 groups according to the so-called Denver classification. Each of the groups comprises all chromosomes of similar size and morphology, and distinguishable from those of all other groups. The two X-chromosomes cannot be distinguished from those of group C.

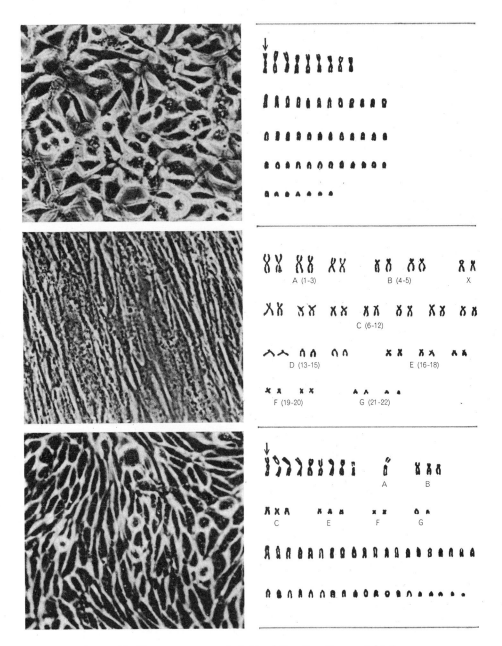

FIGURE 3.1. Human × mouse hybrids: *left*, cell cultures; *right*, karyograms of the mouse parent (*top*), the human parent (*middle*) and the hybrid (*bottom*). From Ephrussi and Weiss[72] with the permission of *Scientific American*.

The comparison of these two karyograms shows that most of the human chromosomes can be distinguished from those of the mouse. The only likely confusions are between the human chromosomes of group A and the long metacentrics of the mouse line; and between those of groups D and G, and mouse telocentrics.

The lower karyogram of Figure 3.1 is of a young hybrid (*ca.* 20 generations old) from this human $\times$ mouse cross. It reveals the presence of 14 human chromosomes, all of which are distinguishable from those of the mouse.

The microphotograph of Figure 3.2 shows a metaphase of a cell of another, older hybrid, clone. In addition to most of the mouse chromosomes, this metaphase contains only three human ones. Since this clone was grown continuously in the selective medium (HAT), in which the mouse (Clone 1D) cells cannot grow because

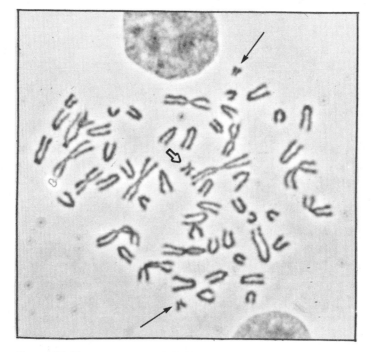

FIGURE 3.2. Human $\times$ mouse hybrid cell which has lost all but three human chromosomes. The black arrows indicate human chromosomes of Group G and the white arrow a human chromosome of Group E (17 or 18) which must carry the TK$^+$ gene. From Ephrussi and Weiss,[72] with the permission of *Scientific American*.

they lack thymidine kinase, at least one of these three human chromosomes must carry the human TK+ gene necessary for growth in HAT medium.

A culture of this clone was therefore transferred to and grown in normal medium supplemented with BUdR which selects *against* cells possessing thymidine kinase activity. This rapidly resulted in selection of cells which had lost a human chromosome and, simultaneously, the ability to grow in HAT medium. This has been observed to happen in clones of several similar hybrids.[163, 167] The relevant chromosome was first thought to belong to group C, but was later identified[163, 167] as a member of group E (probably No. 17 or 18), thus providing strong evidence that the TK+ gene is located in one of these chromosomes. I say "strong evidence" rather than "proof" because the unambiguous proof requires the demonstration that the thymidine kinase activity in these hybrid cells, present at the outset, was indeed due to the presence of a *human* enzyme. As you will see, this proof was provided later.

Weiss and Green also tested several of their clones containing different numbers of human chromosomes, for the presence of human surface antigens (detected by mixed agglutination with human red blood cells in the presence of rabbit antiserum against WI-38 cells). These tests showed that the agglutinability of the different hybrids is roughly proportional to the number of human chromosomes they contain, and that the presence of only very few human chromosomes suffices to give a weakly positive test (Figure 3.3). Aside from providing a further proof of the functional state of the human chromosomes in the hybrid cells, these results indicate that the genes specifying surface antigens are numerous and widely distributed among the human chromosomes.[259]

The experiments on thymidine kinase resulted in the first localization of a human gene on one of the chromosomes of a small group using somatic hybridization. You will have noticed that it was facilitated by the possibility, in this case, to select *for* as well as *against* the retention of a given gene—here, the gene specifying thymidine kinase. (Notice that if two linked markers were present, linkage would be detected by this method.) This is the most direct,[79] but not the only possible procedure for assigning genes to definite chromosomes or chromosome groups. In fact, there are by now two different approaches which I shall briefly describe.

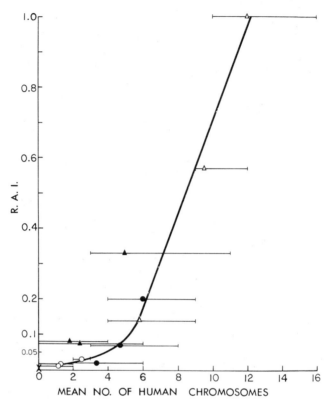

FIGURE 3.3. Relative agglutination index (R.A.I.) as a function of the number of human chromosomes in a series of human × mouse hybrids. From Weiss and Green,[259] with permission.

Two other approaches to the study of linkage

One of these approaches has been initiated by Bodmer and co-workers[169, 182] at Stanford using, to begin with, hybrids between an 8-azaguanine resistant (i.e. HGPRT deficient) mouse cell line and normal human lymphocytes. Selection in HAT medium of hybrids between these two types of cells is especially effective because even though the lymphocytes carry no selective markers, they are nondividing cells.* First it was shown that in these hybrids, the initial, rapid loss of human chromosomes is apparently random, so that different, independently formed hybrids lose different assortments of chromosomes. This phase is followed

* We shall see in Chapter 5 that lymphocyte nuclei are reactivated following fusion with dividing cells.

by a phase of further much slower chromosome loss or stabilization. Since many homologous enzymes of mouse and man can be distinguished by electrophoresis, the presence, in different hybrid clones, of several definite human gene products was correlated with the presence of definite human chromosomes or chromosome groups. Lastly, given the availability of a number of clones having lost different constellations of human chromosomes, whether the different human gene products can be lost independently of each other was established: the constant simultaneous loss of two human enzymes indicates probable genetic linkage (i.e. localization on the same chromosome), while independence of loss indicates absence of linkage (Note 15).

This general approach, first used by Bodmer's group, was soon adopted by others, especially Ruddle and his co-workers[13] at Yale. By the beginning of 1971, it had provided information most of which is summarized in Table 3.1. The following can be seen:

TABLE 3.1. Linkage or absence thereof between human genes as determined by somatic cell hybridization.[a]

Linked genes	Assignment to chromosomes
Hypoxanthine-guanine phosphoribosyl transferase Glucose-6-phosphate dehydrogenase Phospho-glycerate-kinase	X
Lactate dehydrogenase : B subunit Peptidase B	?
Genes unlinked to each other or to the above	
Adenosine deaminase Esterase-2 regulator Indol-phenol oxidase Isocitric dehydrogenase	C_{10}?
Lactate dehydrogenase : A subunit Malate dehydrogenase (NAD-dependent) Phosphoglucomutase 1 Peptidase A Peptidase C	C ?
Thimidine kinase	$E_{17 \text{ or } 18}$

[a] For references, see text pp. 36-38.

1. Of the 15 listed genes, only the first three (HGPRT, G6PD, and PGK) are linked *and* located in the X-chromosome (it may be noted that the sex linkage of these genes was already known from

or indicated by the study of pedigrees[165, 169, 182, 231]): their constant simultaneous presence in segregated hybrids selected in a medium (HAT) where only HGPRT+ cells can grow is therefore not unexpected.

2. Linkage has also been detected[211,215] between the autosomal genes LDH-B and Peptidase B, but they have not been assigned to a definite chromosome.

3. All the other 10 genes are unlinked to each other or to the genes of the two former groups.[13, 169, 182, 211, 215] This means that, taken together, these 15 genes represent 12 linkage groups, while there are 23 different pairs of chromosomes in man. In fact, my table does not contain some new data obtained by the same method—so the situation at present is even better than that.

Table 3.1 also does not contain the data obtained by a different approach described recently by Kao and Puck.[132] This approach is, I think, worth describing.

Prior work by Puck's group has permitted them, using very elegant techniques, to induce and to isolate a number of simple and double nutritionally deficient mutants from a near diploid line of Chinese hamster cells, designated CHO.[205] This cell line has two advantages: (1) it grows extremely rapidly: its generation time is only 12 hours, instead of approximately 24 of most of other cell lines and strains, including human diploid cells; (2) the number of chromosomes is very low; CHO contains only 20 chromosomes, most of which, according to Kao and Puck, can be distinguished from human chromosomes. Now, Kao and Puck have found that hybrids between cells of this hamster line and fibroblasts recently explanted from a human newborn lose the human chromosomes with extreme rapidity. This is shown in the histograms of Figure 3.4. The upper histogram shows that immediately after Sendaï induced fusion, the hybrid cells contain exactly the "expected" number of chromosomes (66): 20 from the hamster, and 46 from the human parent (it is rather regrettable though that the number of analysed metaphases is given as a percent rather than as an absolute number!). Further, it can be seen that after only one week of growth (which at most represents 14 cell generations!) the hybrids have lost a major part of the initial chromosome complement—according to the authors, the majority of chromosomes lost are human ones. Lastly, the histo-

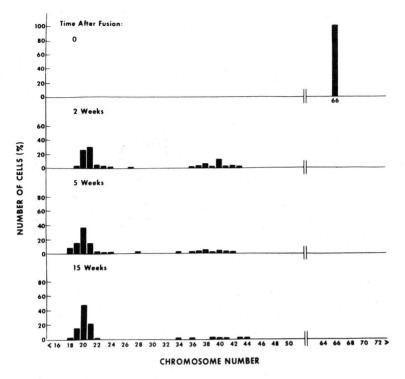

FIGURE 3.4. Distribution of chromosome numbers in uncloned cultures of human × Chinese hamster cell hybrids, at various periods of incubation following fusion of the cells. From Kao and Puck,[132] with permission of *Nature*.

gram shows stabilization of the karyotype at a modal number of 19 chromosomes—1 fewer than the modal number in the CHO line.

The principle of Kao and Puck's experiments on linkage of human genes is as follows. CHO cells with *two* nutritional deficiencies are fused with normal human diploid cells, and loss of chromosomes is allowed to take place in a medium containing *only one* of the required metabolites. Under these conditions, it is clear that if the hybrid clones which grow up are always compensated for *both* nutritional deficiencies of the hamster parent, both normal alleles of the two genes causing the double auxotrophy must be carried on the same human chromosome, provided that the different hybrid clones have only one human chromosome in common. If, on the contrary, the unselected

39

marker were lost randomly, this would clearly show that this marker and the one selected for are not linked, i.e. are located on different human chromosomes.

In practice, the first experiments[132] were performed as follows. CHO cells doubly deficient either for inositol and proline (*ino−pro−*) or for glycine and proline (*gly− pro−*) were mixed with normal human cells (*ino+ pro+*).* The mixed cell suspension was treated with inactivated Sendaï virus and then inoculated into petri dishes in a medium which (1) was supplemented with *dialysed* fetal calf serum, in which the human cells do not grow; (2) contained proline, but no inositol in the first case, and no glycine in the second. After two weeks, the hybrid clones had grown up and were found to have reached the stage of practically maximum chromosome loss. They were then tested for the presence or absence of the original proline requirement. The results are shown in Table 3.2. It will be seen (last column) that

TABLE 3.2. Test for linkage between human inositol and proline and between glycine and proline genes.[a]

Parental genotypes		Proline containing growth media for hybrids	Number of hybrid clones with indicated genotypes	
Human fibroblasts	CHO			
ino+ pro+	ino− pro−	Inositol free	ino+ pro−	35
			ino+ pro+	7
gly+ pro+	gly− pro−	Glycine free	gly+ pro−	10
			gly+ pro+	5

[a] From Kao and Puck[132] (rearranged).

both *ino+* and *gly+* hybrids can be either *pro+* or *pro−*: in other words, there is no close linkage between inositol and proline on the one hand, or between glycine and proline on the other.

This conclusion was confirmed by the reverse experiment[132] in which hybrids between hamster cells deficient for both inositol and proline and normal human cells were selected for inositol sufficiency. Three of the four clones obtained were *ino+ pro−*, one other was *ino+ pro+*.

Figures 3.5A, B, and C show, respectively, the karyotypes of

* In most cases, the exact location of the blocks in the corresponding biosynthetic pathways (i.e. the nature of the missing enzymes) is unknown.

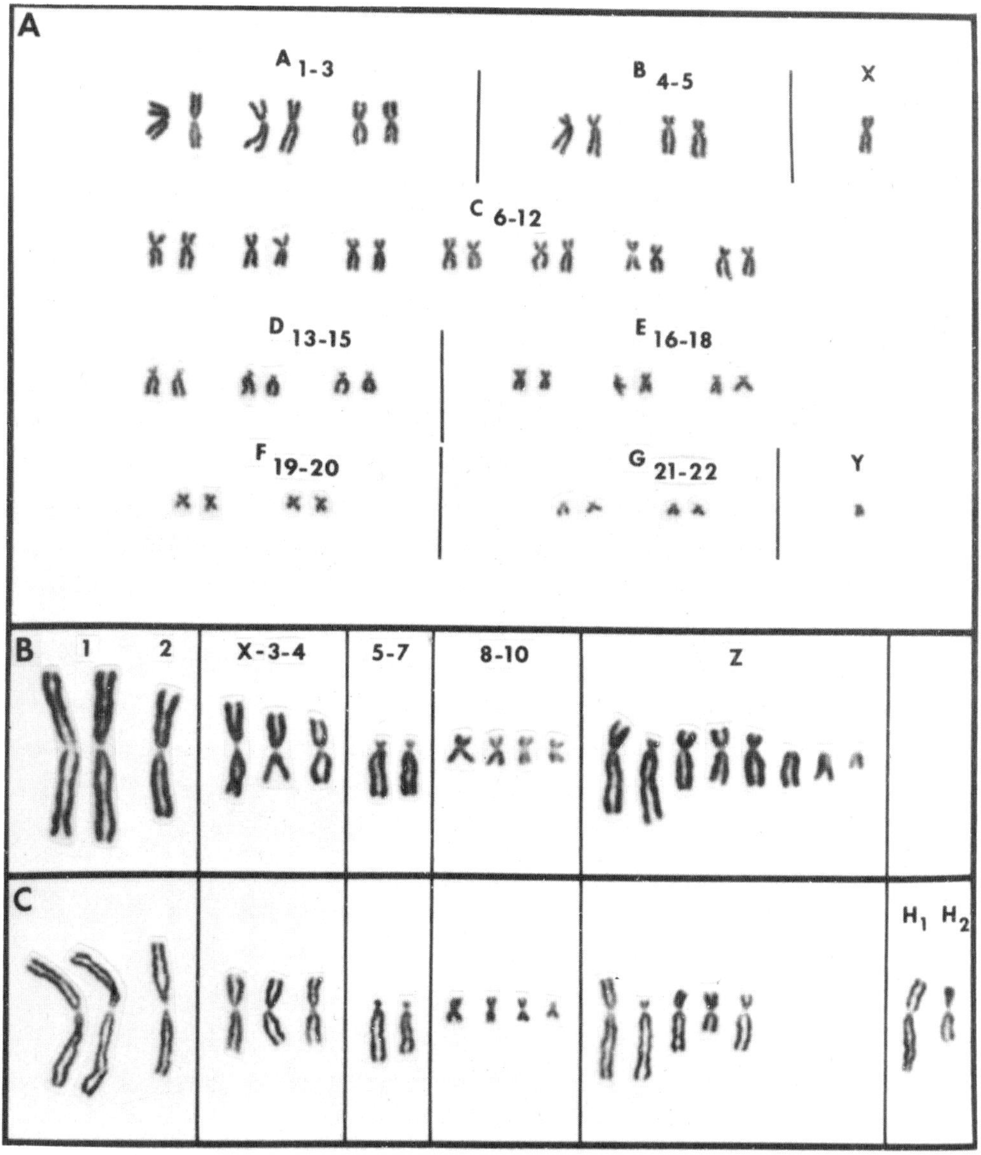

FIGURE 3.5. Karyotypes of parental cells and hybrid of the cross between human fibroblasts and cells of the Chinese hamster line CHO. *A*, karyotype of human fibroblasts; *B*, karyotype of CHO; *C*, karyotype of the ino⁺ pro⁻ hybrid obtained by fusion of A and B and subsequent growth in medium lacking inositol but containing proline. From Kao and Puck,[132] with permission of *Nature*.

the human and hamster cells used in one of the crosses and of one of the "segregated" hybrids. The latter appears to contain only two human chromosomes. This hybrid was *ino+ pro-*. Electrophoresis of extracts of hybrids of this clone and of the parental cells gave the patterns reproduced in Figure 3.6. It will

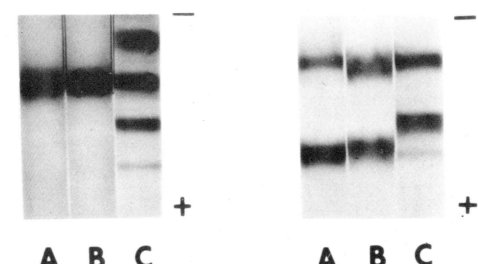

FIGURE 3.6. Gel electrophoretograms of lactate dehydrogenase (LDH) *left* and malate dehydrogenase (MDH) *right* obtained from extracts of: *A*, CHO cells; *B*, the ino+ pro- hybrids; *C*, normal human fibroblasts. From Kao and Puck,[132] with permission of *Nature*.

be seen that both human LDH and human MDH are missing from the hybrids and that the patterns obtained are identical with those of the hamster line. Since the hybrids have *lost* both human LDH and MDH, and *retained* the human inositol+ gene, it is concluded that there is no linkage between the genes for inositol, LDH, and MDH.

I shall only add that in their (preliminary) paper, Kao and Puck do not speak about having tested the segregated hybrids for any other of the 15 enzymes which were listed in Table 3.1, which I compiled chiefly from Bodmer's and Ruddle's data; therefore I could not lengthen the list by adding Kao and Puck's data. This will no doubt be done soon.

The problem of mapping human chromosomes

Be this as it may, my purpose was not to give you the latest on human linkage groups, but rather to show you that we now have the tools for the *first step* in mapping of the human genome: the establishment of linkage groups. It is obvious that further progress towards this first goal will depend as much on the patient joint efforts of cell geneticists in the search of (induced or inborn) mutants of mouse, hamster, and man (especially of those mutants which can be used as selective markers), as on the ingenuity of biochemists and cytochemists in the design of new selective systems and of simple techniques for the cytochemical identification of numerous specific proteins of these species. I have no doubt that a concerted effort of this sort could, before very long, provide us with quite richly marked human linkage groups.

But the next step, mapping of chromosomes, takes more than just assigning genes to linkage groups and chromosomes: as you know, it demands also the definition of the positions of the different linked genes relative to each other. In Drosophila, this was obtained chiefly by the study of the results of crossing over occurring during meiosis. In Drosophila, as well as in fungi, crossing over occurs also in the course of mitosis; but unfortunately, for some unknown reason, the occurrence of mitotic crossing over in mammals has not as yet been observed with certainty. It is indeed possible that it does not occur. Are we then to abandon all hope for eventually mapping human chromosomes?

Fortunately, other mechanisms for mitotic recombination of linked genes, such as translocations, exist and apparently occur in mammalian somatic cells from time to time.

Translocations of genetic material

Let me briefly describe some recent observations made by Barbara Migeon and co-workers,[167] at Johns Hopkins Medical School, on hybrids between cells of Clone 1D (TK⁻) and human diploid fibroblasts from a patient with the Lesch-Nyhan syndrome, known to be associated with lack of HGPRT.

Eight clones of such hybrids were obtained and maintained for 6 months in HAT medium. At the outset, the karyotypes of all eight clones comprised all chromosomes of the mouse parent

43

and one human chromosome of group E: there was little doubt that this was the chromosome carrying the TK+ gene necessary for the survival of the hybrid cells in HAT. However, after 4 months, this human chromosome unexpectedly disappeared in some of the hybrid clones. These clones nevertheless retained thymidine kinase activity, as was shown by incorporation of tritiated thymidine. Although Clone 1D has never been observed to revert to the TK+ condition, the question arose whether the thymidine kinase activity of the hybrids which had lost their only human chromosome was indeed due to the synthesis of the *human* enzyme. Fortunately, human and mouse thymidine kinases can be distinguished by two criteria: (1) electrophoretic mobility and (2) heat sensitivity; the results obtained by both methods concurred in showing that the enzyme synthesized by the hybrid clones, which had no detectable human chromosomes, was indeed of the human type.[168] The results of electrophoresis (in fact, a combination of electrophoresis and autoradiography) are shown in Figure 3.7A, B, and C. It will be seen, by comparing the different slots that the enzyme extracted from various human cells moves (cathodically) faster than that from mouse cells; and that the enzyme extracted from the hybrid cells moves like the human enzyme. It is clear that the hybrids which no longer contain the E-chromosome, contain thymidine kinase of the *human* type and this makes it very plausible to suppose that what occurred in these hybrids was a translocation of a small piece (a piece small enough to be cytologically undetectable) of the human E-chromosome carrying the TK+ gene, into one of the mouse chromosomes.

Another set of observations which *may be* ascribed to the occurrence of translocations was recently described by Miller *et al.*[170] These authors performed several crosses between HGPRT deficient mouse cells of Littlefield's line A-9 and human cells of various types. Since survival in HAT of hybrids from this cross depends on the retention of the human, sex-linked HGPRT gene, it was to be expected that other markers of the X-chromosome will be retained also. It will be recalled that this is what was observed indeed by Bodmer and co-workers (see above p. 37). However, experiments by Miller *et al.*[170] showed that, of 6 segregated hybrids which contained human HGPRT, 2 *had lost human and retained mouse* G6PD. Four others contained both human and mouse G6PD. By re-cloning the latter in HAT, subclones were isolated which contained only the mouse enzyme or both the human and mouse G6PD (Figure 3.8). The authors ascribe these findings to the occurrence in these hybrids of rather frequent breakage of the human X-chromosome, fol-

44

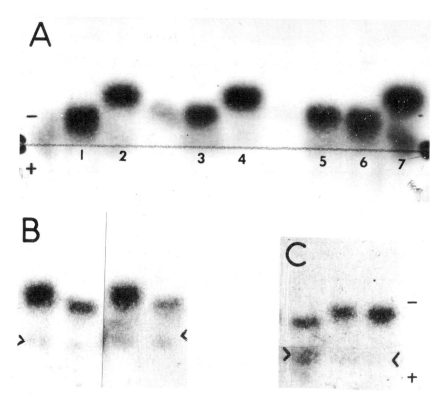

FIGURE 3.7. Starch gel electrophoresis of thymidine kinase from mouse, human, and hybrid cells. *A*, slots 2, 4, and 7, human fetal liver; slots 1 and 6, mouse cells of line L-929; slots 3 and 5, mouse cells of line A-9. *B*, left to right: hybrid, mouse (A-9), in duplicate. *C*, left to right: mouse (A-9), hybrid, human (WI-38). From Migeon *et al.*,[168] with permission of the Plenum Publishing Corporation.

lowed by the retention in every clone of the HGPRT gene, indispensable for growth in HAT, and by the loss in some of the clones of the (apparently loosely linked) nonessential G6PD gene.

Owing to the unfavorable karyotype of A-9 cells, these conclusions could not be substantiated by karyological analysis. It thus remains uncertain whether the breakage of the X-chromosome resulted (1) in the formation of a slightly deleted (shorter) X-chromosome, or (2) in the loss of most of the X-chromosome with translocation of the remaining segment (carrying the retained marker) into a mouse chromosome.

Now, assume that the human linkage groups were as well-marked as those of Drosophila: it is highly likely that we would detect in Migeon's hybrids not only the human TK gene, but also its closest neighbors, for it is not likely that it was translocated

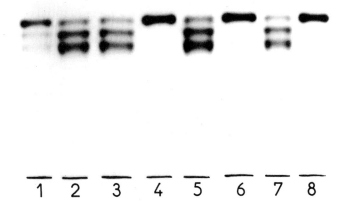

FIGURE 3.8. G6PD isozymes of human × mouse hybrids. Channels 3 to 8 show the enzyme patterns of subclones derived from the hybrid whose pattern is shown in channel 1. Only the mouse enzyme is present in channels 4, 6, and 8. The three-banded pattern in the other channels comprises mouse (top band), hybrid (middle), and human (bottom) enzymes. From Miller *et al.*,[170] with permission.

alone. Assume further that these translocations are of different lengths: we could then begin to define the relative positions of the neighbors of the TK gene, with respect to TK, by the frequency with which they accompany this gene in the different translocations.

The following observations will show that this suggestion is no longer as far-fetched as it first looked.

Schwartz *et al.*[222] have recently described what may be an extreme case of incorporation into somatic cells of what must be very small pieces of foreign genetic material carrying the information for the synthesis of an essential enzyme. In these experiments, mouse fibroblasts of the HGPRT deficient A-9 line were fused with normal chick erythrocytes and cells capable of growth in HAT were obtained. Karyological analysis of these cells provided no evidence for the presence of chick chromosomes; and immunological tests (performed with Watkins and Grace's immune haemadsorption technique[251]) detected no chick surface antigens which are likely to be distributed, like in human cells,[259] among numerous chromosomes. However, the electrophoretic mobility of the HGPRT formed by these cells was found to be clearly similar to that of the enzyme from chick cells and very different from that of mouse cells (Figure 3.9).

Concerning the mechanism of the incorporation of the chick genes into the mouse cells, it is suggested that it starts with the "premature chromosome condensation"[125], [127] (also described as "chromosome pul-

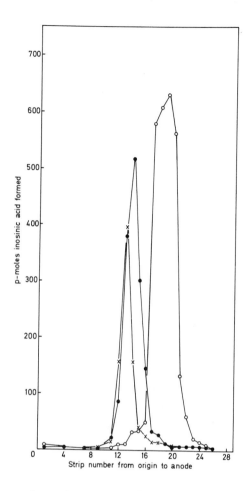

FIGURE 3.9. Electrophoretic mobility of HGPRT produced by mouse fibroblasts of line A-9, chick erythrocytes, and hybrids between them. O, A-9 cells; ●, chick erythrocytes; ×, hybrids. From Schwartz et al.,[222] with permission of Nature.

verization"[133, 134]) which has been observed to occur when mitotic (HeLa or A-9) cells are fused with interphase cells of the same or different species.[125-127] Examples of this phenomenon are given in Figure 3.10. Schwartz et al.[222] suggest that when the chick erythrocyte + A-9 heterokaryons enter mitosis, the erythrocyte chromosomes are fragmented, and that fragments carrying the HGPRT gene, indispensable for growth in HAT, are retained. As to their exact cellular localization in the "hybrid" cells (i.e. their mode of integration), it appears probable, at first sight, that they are translocated into mouse chromosomes. If this were so, one would expect that the chick (HGPRT) genes should not be frequently lost from these cells upon growth in normal (nonselective) medium. In fact, however, their loss appears to be much more frequent than expected, indicating a rather loose mode of (nuclear?) integration.[222]

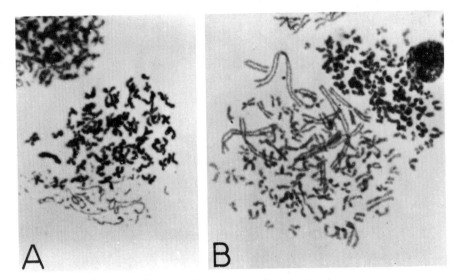
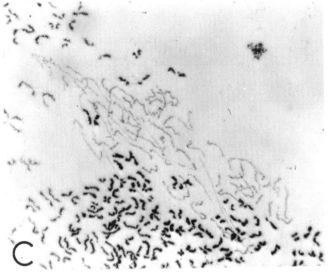

FIGURE 3.10. Premature chromosome condensation (PCC) following fusion of mitotic HeLa cells with various other cells in different stages of the cell cycle. *A*, G1 type of PCC of the embryonic chick erythrocyte nucleus in a HeLa + chick erythrocyte heterokaryon; *B*, G2 type of PCC of the nucleus of a Chinese hamster (CHO) nucleus after fusion of a CHO cell with a mitotic HeLa cell (the metaphase HeLa chromosomes are more condensed than those of the CHO cell in G2); *C*, PCC in a HeLa + horse lymphocyte heterokaryon. Note the single stranded G1 chromosomes of the horse lymphocyte by comparison with the metaphase HeLa chromosomes (and with the double stranded CHO chromosomes in *B*). From Johnson *et al.*,[127] with permission of the Wistar Institute Press.

It may appear to you that I am taking an unduly optimistic stand and that, in any case, work ahead of us is endless and fastidious. I think my optimism is justified by the recent rapid progress in the field, as well as by the acceleration one may expect from some as yet unpublished, technical tricks about which I unfortunately have no time to speak (Notes 14 and 16). As to the dullness of the work in formal genetics, I shall not say that I disagree. However, whether we like it or not, it is absolutely indispensable for the understanding of any other problem of cell biology, such as that of cell differentiation which, whatever the pessimists say, still represents a major unsolved problem to which I shall turn now.

CHAPTER 4

Preliminary Technical and Theoretical Remarks on the Study of Cell Differentiation *In Vitro*

As far as development is concerned there are still things [the enquiring scientist] wants to know. . . . We might predict that the conclusions will never be of the magnitude of Mendel's laws or Darwin's theory of natural selection; the generalization of that importance lies in the epigenetic theory of development. But there is ample room for lesser generalizations within this framework.

(John Tyler Bonner. 1960, p. 518)

Technical remarks

I have shown earlier that we have the means for beginning the genetic analysis of somatic cells of higher animals; what I have to show now is that we also have at least *some* material to which the genetic method can be applied for the purpose with which we shall be concerned first, i.e. cell differentiation.

As I have mentioned already, Harris and his co-workers have shown that even highly differentiated cells such as macrophages, lymphocytes, and erythrocytes, which normally do not divide anymore, can be fused with dividing cells as well as with each other.[98, 107] On the other hand, the successful cultivation of cells maintaining over long periods of time the characteristics of their original differentiation has by now been achieved in many cases (for review, see Ref. 82). I must, however, qualify this statement. What we have at our disposal today are mostly permanent lines of neoplastic cells derived from differentiated *tumors*: for example, hormone producing pituitary and adrenal tumor cells,[216, 245] melanomas which continue to produce melanin,[116, 176] or hepatomas which go on producing some liver specific proteins.[199] Since it may be argued that the stability of differentiated functions in these cell lines is exceptional, and due precisely to the altogether abnormal regulation of these cells, correlated with their neoplasticity, I hasten to add that the refinements of cell culture techniques have recently revealed that, under certain conditions, normal differentiated cells can also be propagated *in vitro* for long periods of time without loosing the differentiated functions characteristic of the tissue of origin. This has been demonstrated, in particular, for chick and rat muscle cells by Konigsberg[141] and Yaffé;[209, 265] for chick cartilage cells by Coon;[32] for chick retinal pigment cells by Cahn and Cahn;[23] and it has been shown that these cells, under appropriate conditions, continue the synthesis of their characteristic differentiated products: muscle fibers (i.e. actin and myosin), chondroitin sulfate, and melanin, respectively.* Figure 4.1A gives an example of the expression of a differentiated function under condition of growth *in vitro*. "Dedifferentiation," i.e. the total and irreversible loss of specialized functions, long believed to be the inevitable result of

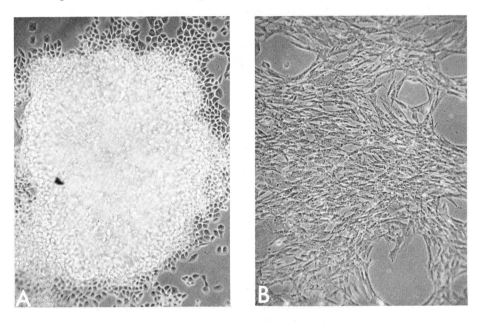

FIGURE 4.1. Clones of cartilage cells grown in two different media. The cells of the colony shown in *A* are embedded in a metachromatic matrix. *B* shows a sibling clone growing in a different medium; note the fibroblast-like growth habit of the cells and the absence of metachromatic matrix. From Coon,[32] with permission.

* Additional references on this point will be found in the "Discussion" of the paper by Pfeiffer *et al.*[196]

growth *in vitro*, is sometimes observed; but under appropriate conditions is neither rapid nor inevitable.

Determination and differentiation

I hope I will be excused if, before I tell you about crosses involving differentiated cells, I make a certain number of theoretical statements, some of which are definitely controversial.

First, the ability of serially propagated cells to exhibit *in vitro*, over many cell generations, differentiated tissue specific functions, taken together with the demonstration (accepted, I think, by everybody) that the nuclei of all cells of higher animals contain the complete genetic information of the species, leads one to conclude that something underlying each type of differentiation is *heritable*. That it is not the overtly differentiated state *itself* which is inherited, is clearly shown by the fact that manipulation of the medium can often result in the disappearance of visible differentiation; yet the latter (i.e. the *same* differentiation) will be re-expressed upon re-establishment of adequate conditions (Compare Figures 4.1A and B). Thus, what is inherited is not the differentiated state as such, but the ability to undergo a certain type of differentiation, or, if you prefer, *the commitment to a certain course, to the exclusion of all others*. This commitment is the result of what embryologists call "determination," and it is the *determined state*, and not the *differentiated state*, which is inherited (Note 17). Changes of commitment *are* known to occur, but they are rare. The embryologists call this event "transdetermination"[94, 263] and the pathologists "metaplasia."[74, 160]*

I am fully aware of the fact that the distinction I am making between determination and differentiation is not accepted by everybody. Thus, in a recent review entitled "Biochemistry of differentiation," P. Gross[87] states: ". . . There is no evidence, experimental or logical, in favor of the idea that commitment to a course of change involves molecular mechanisms distinct from the final realization of that change." Let it then be, on my part, a first article of faith (I shall soon come to some others!), and let me add that molecular biologists as competent as Brown and

---

* In metaplasia, the activity of a cell of a given tissue is changed to that typical of cells of another tissue. A good example of this is the secretion by certain pulmonary tumors of a hormone normally produced by the parathyroid gland.[223]

Dawid[17] (not to mention embryologists like Abercrombie[1]) share this heresy.

I am, of course, fully aware that the degree of stability of determination varies tremendously through the animal kingdom. The significant thing to me though is precisely that it is so variable and that it *can be* so extraordinarily stable.

Household and luxury functions

Whether this distinction between determination and differentiation is valid or not, there is a second distinction which I like to make and which must be regarded as another of my articles of faith, since it is not accepted by everybody either. This distinction is between *ubiquitous* functions, i.e. metabolic functions essential for the maintenance and growth of *any* cell, and *differentiated* functions which must be necessary for the survival of the multicellular organism and of the species, but not for that of the cell. For many years I have spoken of the gene products performing these two sorts of functions as "household items" and "luxury items," respectively, and I am glad to see that this terminology has by now been adopted by several distinguished authors, in particular H. Holtzer (cf. Ref. 114) (Note 18).

Putting the two things together, I shall then say that a determined cell, like any cell, produces all necessary household items; but when it begins to differentiate *and only then*, it produces, in addition, a set of luxury items characteristic of the particular cell type.

The epigenotype and the program of development

Turning from the different proteins produced by different types of cells, to the genes which specify their structure, I shall use the term "epigenotype" recently proposed by Abercrombie[1] to designate that part of the total genome which, under appropriate conditions, *can be* expressed in a given cell type *to the exclusion of those limited to other cell types*. Using this terminology, we can then say that different cell types have different epigenotypes (Note 19).

To what these different epigenotypes correspond at the cellular or molecular level is open to question and indeed is the subject of lively speculations. (See Ref. 87.) Aware of the fact that differ-

53

entiation is readily reversible in some (but not all) forms and above all, of the ability of amphibian somatic nuclei to support the normal development of enucleated eggs demonstrated by nuclear transplantation,[88, 91, 147] most people concerned agree, I think, that differences in epigenotype are not due to truly genetic changes (in the sense that gene mutations are). They therefore consider different epigenetic states as corresponding to different regulatory states resulting in the expression of different parts of the total genetic information in different cell types. However, there is no agreement as yet as to whether this is due to selective transcription of the genome; to selective destruction of messenger RNA; to selective transport of messenger into the cytoplasm; to selective translation; or to selective gene amplification[87] [a process which is gaining popularity although in so far as I am aware, it has been definitely demonstrated thus far only for the cistrons responsible for the synthesis of ribosomal RNA (see Ref. 17 and Note 20)]. I shall add that relatively little consideration is given to regulation at other levels: there is, however, increasing and, I think, justified interest in the regulatory role of the cell membrane.

Concerning the stability of the differentiated state, I think it is fair to say that most molecular biologists think of it in terms of transcriptional regulation of the bacterial or phage type, and tend to ascribe it to self-maintaining regulatory circuits; and that some others think of stable restrictions on transcription by self-perpetuating association of chromosomal DNA with histones and/or acidic proteins.*

I must mention at this point that the validity of the evidence for cell heredity in clones of determined or differentiated cells (i.e. for the perpetuation of *intrinsic* differences between cells of different histo-genetic types) is altogether contested by Grobstein.[84]

Grobstein[86] briefly states his position in the following quotation from a recent discussion of the subject: "None of the experiments said to indicate that differentiative instruction is replicable through hundreds of generations shows that this involves a mechanism at the level of the individual cell. They are all experiments on multicellular systems, where the possibility of cell-to-cell interaction as a stabilizing factor always

* For interesting speculations and additional references, see the recent papers by Britten and Davidson[16] and Tsanev and Sendov,[248] and the review by Gurdon and Woodland.[93]

exists. The experiment that would satisfy me that we need to postulate stability at the level of the cell and its genome is one in which daughter cells are separated and cultured separately at every successive cell generation for say, ten generations. If differentiated properties persist in these conditions we must look for a cellular heredity system to explain it."

Commenting on Grobstein's views, Abercrombie[1] writes: "It seems to be generally felt, however, in spite of the convincing evidence that Grobstein rightly demands, that the marked ability of vertebrate cells to maintain a replicated difference when populations of them are exposed to a range of common environments, *in vivo* and *in vitro*, justifies the hypothesis that individual cells carry intrinsic replicated differences independently of their neighbours."

From what I have said earlier about cell heredity, it must be clear to you that I do not share Grobstein's view and believe that some of the experiments I shall describe support my position.*

As to my own feelings about all this, I shall try to sum them up in a few sentences.

First, my impression is, I am afraid, that most of the hypotheses I mentioned are concerned with differentiation rather than with determination; and that much of what is considered as a cause of differentiation is the effect of the still completely mysterious and elusive act of determination.

Second, I think that if what Hershey[110] calls the *unwritten dogma* is correct (i.e., "the inference that *all* three-dimensional structure is encoded in nucleotide sequences"), the *establishment* of different epigenotypes in the course of development must be coded for nuclear DNA because the whole program of development is transmitted from generation to generation. But I also think that whether the functional restriction of the total information, which results in different epigenotypes of different cell lineages, is due to a change in the chromosomes themselves (as it seems to be in the inactivation of one of the Xs in mammalian females, for example; see Note 21) or is only a *reflection* of a change elsewhere in the cell (say, in the cell membrane) is an entirely separate and largely unresolved question worthy of very serious consideration and to which I shall return in Chapter 8. In fact, this is *the* fundamental question to which I have no answer.

* What I have in mind are the observations on the stability of epigenetic changes and on their re-expression which will be described in Chapter 6 and discussed in Chapter 8.

I do believe, however, that unexpected solutions to this and other problems I have outlined may emerge from any of the approaches which have led to the formulation of the theories I have—very incompletely—reviewed. I trust that the genetic approach, for which cell hybridization is at present the only available tool, may prove to be particularly rewarding in this respect.

CHAPTER 5

Application of Cell Hybridization to the Study of Differentiation:
I. Reactivation of Nuclei of Differentiated Cells in Heterokaryons

In short, the whole process of information transfer can be analysed in these heterokaryons with a degree of precision not attainable in any other biological system.

(Henry Harris. 1970, p. 65)

I shall now speak of some insights into the problem of cell differentiation gained by hybridization of somatic cells, and I shall begin with the work of Henry Harris' group on heterokaryons involving rabbit macrophages, rat lymphocytes, and hen erythrocytes. All three of these cell types represent terminal stages of differentiation: normally, none of them synthesize DNA. Erythrocytes, which *are* nucleated in birds, are the most specialized of the three types of cells; while macrophages and lymphocytes synthesize RNA, erythrocytes apparently do not,* or only in minute amounts undetectable by the methods used by Harris and his coworkers.[98, 99, 107]

As I told you earlier, these three types of differentiated cells have been fused both with actively dividing human and mouse cells and with each other, and you have seen what some of these heterokaryons look like (Figures 2.9 and 2.10).

Synthesis of RNA and DNA in heterokaryons

Having ascertained that it is possible to produce such combinations, Harris and his co-workers began with an autoradiographic study of RNA and DNA synthesis† by the nuclei of the three differentiated cell types (mentioned above) in the various heterokaryons. Their results are summarized in Table 5.1. The examina-

* See however Ref. 128.
† Using tritiated uridine and thymidine.

TABLE 5.1. Synthesis of RNA and DNA in heterokaryons.[a]

	RNA[b]	DNA[b]
Cell type		
HeLa	+	+
Rabbit macrophage	+	0
Rat lymphocyte	+	0
Hen erythrocyte	0	0
Cell combination in heterokaryon		
HeLa — HeLa	++	++
HeLa — rabbit macrophage	++	++
HeLa — rat lymphocyte	++	++
HeLa — hen erythrocyte	++	++
Rabbit macrophage — rabbit macrophage	++	00
Rabbit macrophage — rat lymphocyte	++	00
Rabbit macrophage — hen erythrocyte	++	00

[a] From Harris et al.[107]

[b] 0, no synthesis in any nuclei; 00, no synthesis in any nuclei of either type; +, synthesis in some or all nuclei; ++, synthesis in some or all nuclei of both types.

tion of this table permits deduction of the following rules, which I quote from Harris et al.[107]

"(1) If either of the parent cells normally synthesizes RNA, then RNA synthesis will take place in both types of nuclei in the heterokaryon, even if one of the parent cells normally does not synthesize RNA. (2) If either of the parent cells normally synthesizes DNA, then DNA synthesis will take place in both types of nuclei in the heterokaryon, even if one of the parent cells normally does not synthesize DNA. (3) If neither of the parent cells synthesize DNA, DNA synthesis will not take place in the heterokaryon."* And further: "It will be noticed that in all cases where a cell which synthesizes a particular nucleic acid is fused with one which does not, the active cell initiates the synthesis of this nucleic acid in the inactive partner. In no case does the inactive cell suppress synthesis in the active partner."

I may add that these rules apparently prevail *whatever* the ratio of the two types of nuclei in the heterokaryon, and whether the active partner is a heteroploid cell of a different animal species or a diploid cell of the same species as the inactive one.

* As already mentioned on p. 22 and in Note 10, HeLa-Ehrlich ascites heterokaryons present a significant exception to rule 2.

Changes in "dormant" erythrocyte nuclei
in heterokaryons with dividing cells

These observations are interesting, especially those involving chick erythrocytes because, as you know, lymphocytes and apparently some macrophages can be induced by certain stimuli to resume DNA synthesis and to multiply, while mature erythrocytes cannot.

Erythrocytes are highly specialized differentiated cells. Although those of adult birds are nucleated, their nuclei are small and "dormant," and have no well formed nucleoli; instead, they contain a number of deeply staining aggregates of DNA and protein called "nuclear bodies" or "chromocenters," clearly visible in Figure 5.1A.

An additional interesting feature of these red cells is that their reaction to Sendaï virus undergoes changes in the course of their maturation. Mature erythrocytes from *adult* hens are lysed by the virus and converted into "ghosts," so that they contribute to the heterokaryons essentially only their nuclei and the cell membrane with very little, if any cytoplasm.[98, 101, 219] The same holds for their immature but nondividing precursors, reticulocytes of the definitive series (obtained from 12–15 day-old chick embryos), which do synthesize RNA and hemoglobin but very little DNA. Still earlier red cell precursors, the dividing erythroblasts of the primitive series, synthesize DNA, as well as RNA and hemoglobin; they are *not* lysed by Sendaï virus so that they bring into heterokaryons both their nuclear and cytoplasmic contents (Refs. 100, 105 and personal communication of Dr. Peter Cook).

The evolution of mature erythrocyte nuclei during their activation in heterokaryons formed by fusion with HeLa cells can be followed.[99] Figure 5.1B shows the erythrocyte nucleus in a heterokaryon soon after fusion with a HeLa cell: one can still see the nuclear bodies. Further evolution of the erythrocyte nuclei is illustrated in Figures 5.2A and B. You will notice the gradual enlargement of the erythrocyte nucleus (about 20–30 fold in volume) which begins a few hours after fusion. This increase in nuclear volume is accompanied by the gradual disappearance of the nuclear bodies and a decrease of nuclear staining owing to dispersion of the previously condensed chromatin.[99] The amount

59

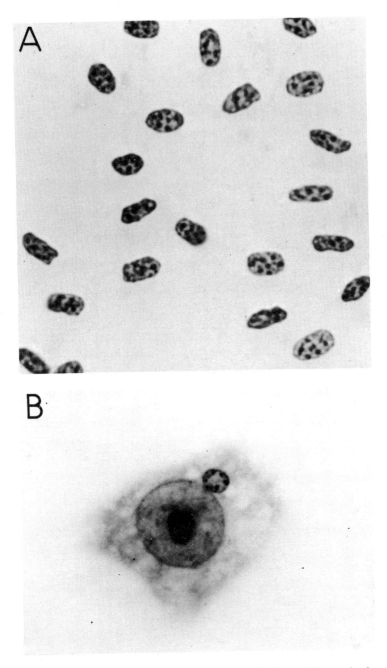

FIGURE 5.1. *A*, Smear of hen erythrocytes in which the nuclei are stained with Weigert's iron haematoxylin; *B*, dikaryon containing one HeLa nucleus and one hen erythrocyte nucleus soon after fusion.[90] Courtesy of Henry Harris.

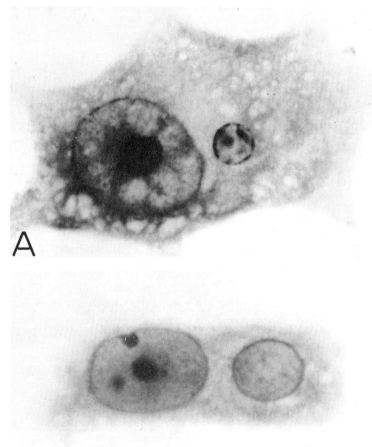

FIGURE 5.2. Further evolution of a HeLa + hen erythrocyte dikaryon. *A*, Enlargement of the erythrocyte nucleus and decrease of staining of the nuclear bodies; *B*, Further increase of the erythrocyte nucleus, disappearance of nuclear bodies.[99] Courtesy of Henry Harris.

of DNA begins to increase almost at once and is preceded by marked changes of the physical properties of the chromatin—the binding of acridine orange increases 4 to 5 fold and the melting profile changes also before DNA synthesis begins.[11] The synthesis of RNA also begins at once, and the amount of RNA synthesized is proportional to the volume of the nucleus.[99]

61

The syntheses of both kinds of nucleic acids can be largely suppressed by irradiating the erythrocytes with UV light prior to the fusion with HeLa or mouse (A-9) cells.[99, 224] In spite of this, the increase in the nuclear volume and the loosening of the chromatin do take place in the heterokaryon, indicating that it is not the synthesis of nucleic acids, but the change in nuclear volume [said to be due to the entry of (foreign!) proteins[101]], which is the primary event resulting in the activation of the "erythrocyte" nucleus.[11, 99] *

Speculations concerning the mechanisms of reactivation of dormant nuclei and regulation of transcription

Harris[101] remarks: "All these observations reflect the fact that nuclear enlargement loosens the chromatin and renders it more accessible, not only to macromolecules, but even to smaller molecules such as the acridine dyes and actinomycin. The same process no doubt also renders the chromatin more accessible to the molecules involved in its transcription, so that, as more of the initially condensed chromatin opens up, more of it is transcribed." Elsewhere (and earlier) he observes:[99] "What we know about the condensation of chromatin in the nuclei of animal cells suggests that it involves rather large areas of the DNA: whole chromosomes or major portions of chromosomes. If this is true, then the control of genetic activity associated with changes in nuclear volume must be a rather coarse regulatory mechanism. Indeed, it has been proposed that transcription of DNA at discrete genetic loci can be prevented by specific cytoplasmic substances (repressors) which are able to recognize these loci and attach to them [Jacob and Monod, 1961†]. There is at present no evidence for the existence of a mechanism of this sort; but if it does exist, its operation in animal cells must be subject to the constraints imposed by changes in nuclear volume: those areas of the chromatin in which the synthesis of RNA has been suppressed by the process of condensation cannot, presumably, be subject to any further regulation."

However, according to Ringertz and Bolund,[210] very similar nuclear changes (increase of acridine orange binding by and change of melting profile of DNA, decondensation of chromatin etc.) can be brought about by washing chick erythrocytes alone, or the nuclei isolated from them, with serum free salt solutions or by exposing these nuclei to agents chelating divalent ions such as EDTA. They also occur at low tempera-

---

* The evaluation of the experiments of Harris' group (described in this Chapter) on the changes which occur in *erythrocyte* nuclei is rendered difficult by the indiscriminate use of this term whether the cells referred to are taken from adult hen or embryos of various stages. Even in the blood of adult birds all erythrocytes are not completely mature; and the "erythrocytes" from embryos of different ages are heterogeneous populations of red cell precursors.[3, 128]

† My Ref. 119.

ture and apparently are not affected by a number of enzyme inhibitors, but are prevented by divalent metal ions.

It has been suggested that the removal of divalent cations from the chromatin may be an essential (nonspecific) factor in the reactivation of dormant nuclei and in the regulation of DNA transcription[11, 101, 210] (Note 22).

Were I to adopt Harris' manner of using references to classics,* I would point out that the discovery of the reactivation of dormant nuclei goes back to 1875 when Oskar Hertwig showed that fertilization results from the activation of the most condensed nucleus of the most specialized cell—the spermatozoon.[111] I prefer to say, however, that it is indeed a remarkable fact that the nuclei of such extremely differentiated cells as mature erythrocytes can be reactivated, and that Harris and his co-workers have taken this opportunity to test what exactly the reactivated erythrocyte nuclei are capable of doing. We shall see that the experiments to be described provide evidence that the process of differentiation which results in the "dormancy" of the red cell nucleus is reversible; and confirm the conclusion from our earliest observations on interspecific hybrids that the signals which regulate cell activities are nonspecies specific.

Reappearance of household functions and the role of the nucleolus in the transfer of genetic information

While the syntheses of DNA and RNA by the erythrocyte nuclei in heterokaryons with dividing cells starts very soon after fusion,[98, 99] the transport of the latter into the cytoplasm, according to Sidebottom and Harris, has to await the development of a full-sized nucleolus.[224] This stage of nuclear reactivation of the erythrocyte nucleus is, however, usually not reached, apparently because, in most of the heterokaryons, the nuclei of the *active* "host cells" undergo within 4 days irregular mitosis and the cells die. More complete reactivation of the erythrocyte nuclei can be obtained if the active cells are given a heavy dose of gamma radiation (6000 rads) prior to fusion.[105, 224] Heterokaryons thus formed survive for much longer periods of time (up to 3 weeks) and within them, the erythrocyte nuclei undergo more complete activation as witnessed by the development of full-sized nucleoli. It is only thereafter that, according to Side-

* See Note 2.

bottom and Harris, the transport to the cytoplasm of the RNA synthesized by the reactivated erythrocyte nuclei begins.[224]

This conclusion finds support in experiments which made use of a technique of irradiation by a microbeam of UV light, which permits the inactivation in a chosen cell either of the whole nucleus, or of only the nucleolus.[224] Sidebottom and Harris thus showed that irradiation of the nucleolus alone of HeLa cells suffices to arrest the transfer to the cytoplasm of RNA synthesized either in the nucleolus itself or elsewhere in the nucleus.[224]

With all the described facts at hand, Harris and associates proceeded to test the reactivated red cell nuclei with respect to synthesis of (1) chicken surface antigens[105] and (2) a specific enzyme (HGPRT).[102] For these studies, heterokaryons were made by fusing chick erythrocytes with irradiated HGPRT deficient mouse fibroblasts of line A-9.

I shall describe first the results of the tests for synthesis of chick antigens performed by means of an ingenious technique of immune hemadsorption worked out by Watkins and Grace.[251]

As mentioned above, if red blood cell precursors from 5-day-old chick embryos are fused with mouse fibroblasts, the former are not lysed by the virus, and the heterokaryons comprise the cytoplasms as well as the nuclei from both cell types (p. 59). The surface antigens of both species are therefore present from the start and are detected as long as the heterokaryons survive (P. R. Cook, personal communication). The situation is different if reticulocytes from 12-day-old embryos or mature erythrocytes from adult hen are employed: in these cases the lysed cells of the red blood series contribute to the heterokaryons little, if any, cytoplasm. The results of tests for the presence of chick antigens provided the results shown in Figure 5.3. This figure gives also the results of measurements of the size of the red cell nuclei, and of determinations of the percent of heterokaryons in which these nuclei are enlarged and contain visible nucleoli. The essential facts which emerge are that right after fusion, chicken antigens are present on 100% of the heterokaryons (curve of black triangles), but that during the next 4 days, this percent goes rapidly down owing to the gradual disappearance of the antigens contributed by the erythrocyte ghost (membrane) and to the absence of chick-antigen synthesis. The latter, however, is resumed on the

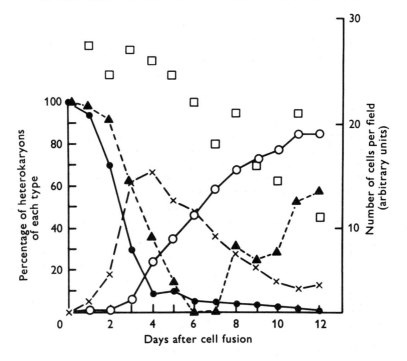

FIGURE 5.3. Reappearance of hen-specific antigens on the surface of heterokaryons made by fusing gamma-irradiated A-9 cells with adult hen erythrocytes. The relationships between the enlargement of the erythrocyte nucleus, the development of nucleoli within it, and the disappearance and reappearance of the hen-specific surface antigens are shown. □, total number of cells; ●, heterokaryons with unenlarged erythrocyte nuclei; ✕, heterokaryons with enlarged erythrocyte nuclei, but not visible nucleoli; O, heterokaryons with enlarged erythrocyte nuclei containing visible nucleoli; ▲, heterokaryons showing hen-specific surface antigens.[105] Courtesy of Henry Harris.

8th day when the curve starts rising again, and this rise, as can be seen in Figure 5.3, is correlated with that of the percent of erythrocyte nuclei having developed full-sized nucleoli.[105]

The reactivation of the red cell nucleus results also in renewed synthesis of chicken HGPRT which is easily demonstrable in this system because, as you will recall, the mouse fibroblasts used in this series of experiments lack HGPRT, the enzyme necessary for the incorporation of hypoxanthine: A-9 fibroblasts exposed to tritiated hypoxanthine show very little radioactivity in autoradiographs. When irradiated fibroblasts were fused with erythrocytes or their precursors, and the heterokaryons thus produced were

tested in this manner for the synthesis of HGPRT, it was found that enzyme activity appeared; indeed, the younger the red cell component, the sooner enzyme activity appeared, and in close correlation with the rate of increase of size of the erythrocyte nucleus and with that of the appearance of a visible nucleolus.
[101, 102]

According to P. R. Cook (personal communication) HGPRT activity appears on the 1st, 4–5th, and 5–6th days in heterokaryons formed by fusion of A-9 fibroblasts with red blood cells from 5-day-old, 12-day-old embryos, and adult hens, respectively. Figure 5.4 illustrates the appearance of the enzyme in heterokaryons of the first type.

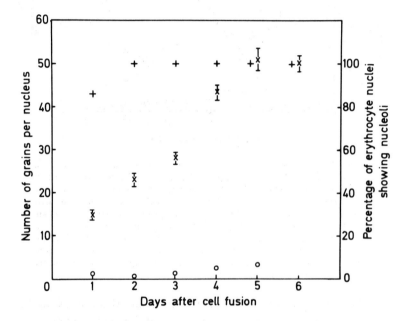

FIGURE 5.4. The development of HGPRT activity in A-9 + 5-day-old chick embryo erythrocyte dikaryons, as measured by their ability to incorporate tritiated hypoxanthine into nucleic acid. The incorporation of hypoxanthine in the dikaryons is initially only marginally greater than that in the A-9 cells, but when the erythrocyte nuclei develop nucleoli, this incorporation increases markedly. O, A-9 cells; ×, dikaryons; +, erythrocyte nuclei showing nucleoli. Courtesy of P. R. Cook (unpublished).

That the enzyme produced by these heterokaryons is of the chicken type, has been clearly demonstrated by Cook.[31] Figure 5.5A shows that, under the conditions used, the electrophoretic

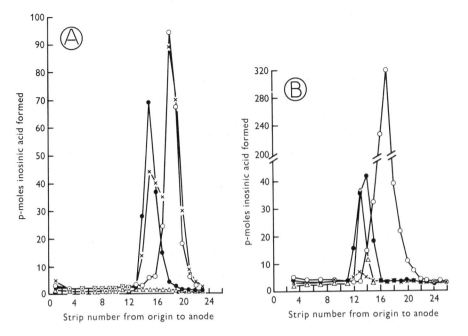

FIGURE 5.5. *A*, Electrophoretic mobilities of HGRPT from mouse and chick cells. O, L-cell extract; △, A-9-cell extract; ●, extract from 12-day-old chick embryo erythrocytes; ×, mixture of extracts from L cells and 12-day-old chick embryo erythrocytes; *B*, electrophoretic mobility of the HGPRT induced in A-9-chick erythrocyte heterokaryons compared with that in L cells and 12-day-old chick embryo erythrocytes. O, L-cell extract; ●, extract of 12-day-old chick embryo erythrocytes; ×, extract of A-9-chick erythrocyte heterokaryons 1 day after cell fusion; △, extract of A-9-chick erythrocyte heterokaryons 9 days after cell fusion. From Cook.[31] Courtesy of Henry Harris.

mobilities of mouse and chick HGPRTs are different, and that the two enzymes can be identified in mixtures of extracts of mouse and chick cells. Figure 5.5B shows that the enzyme activity which appears in heterokaryons on the 1st day after fusion and sharply increases by the 5th day is due to the presence of HGPRT clearly similar, in its electrophoretic mobility, to the chick enzyme.

The luxury function of red blood cells
in heterokaryons with fibroblasts

The reactivation of the chick red cell nucleus thus clearly results in the synthesis of chick ubiquitous (household) products. But what about the synthesis of hemoglobin, the "luxury item," in the production of which the erythrocyte is so highly and uniquely specialized?

According to Harris,[101] no hemoglobin synthesis can be detected in heterokaryons produced by fusion of A-9 fibroblasts with either mature erythrocytes from adult hens or with blood cell populations from 12-day-old embryos, even though the latter comprise chiefly reticulocytes which still synthesize RNA and hemoglobin, and even though, as shown earlier (pp. 63ff.), the reactivated nuclei of both types of red blood cells synthesize surface antigens and HGPRT from the 4th–6th day on. However, a different result is obtained with heterokaryons of the same mouse cells with earlier red cell precursors (primitive, dividing erythroblasts), which are not lysed by Sendaï virus and, hence, contribute to the heterokaryon their cytoplasm as well as their nucleus. Measurements by Cook[101] of hemoglobin synthesis* by such heterokaryons have given the results shown in the curve of Figure 5.6. One can see that there is, immediately after fusion, a *sharp rise* of hemoglobin synthesis, the rate of which may exceed the normal rate by a factor of 10. However, within 24 hours this rise is followed by a rapid *decline* and, after 3 or 4 days, complete *abolition* of hemoglobin synthesis.

I shall not go into the still uncertain interpretation of the differences between the results obtained with more or less mature erythrocytes, on the one hand, and of the rise and fall of hemoglobin synthesis in heterokaryons involving very early red cell precursors (erythroblasts), on the other. The interpretation of all these observations must await further information concerning, in particular, the nature of the hemoglobin produced (chick or mouse?) in the latter case, and the origin of the ribosomes responsible for the hemoglobin synthesis (or rather the Fe^{59} incorporation): is this synthesis assured by pre-existing erythroblast ribosomes, which were already engaged in hemoglobin synthesis, or by newly formed ribosomes?

In his book[101] (pp. 55–56) Harris remarks: "There is as yet no experimental information about the mechanism by which the more gradual, but *permanent, suppression* of haemoglobin synthesis is achieved" [in the heterokaryons involving "early embryonic erythrocytes," i.e. probably mostly erythroblasts, see above p. 59]. "The possibility that

* Estimated by incorporation of Fe^{59}.

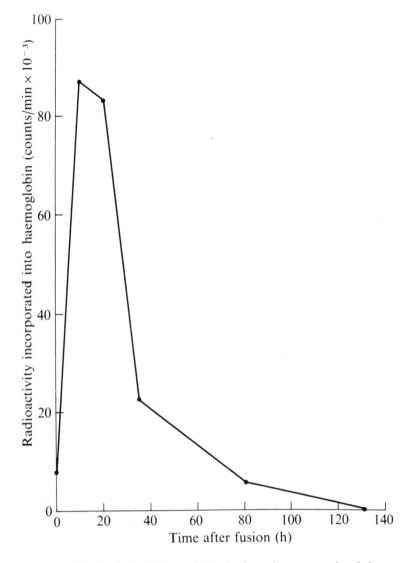

FIGURE 5.6. Synthesis of hemoglobin in heterokaryons produced by fusing immature erythrocytes with mouse tissue culture cells. Note the initial increase in the rate of hemoglobin synthesis rapidly followed by a decrease. The cultures contained about 120 erythrocyte nuclei per 100 cells. From Harris,[101] with permission of Oxford University Press.

this process may involve the destruction of the RNA derived from the erythrocyte is being explored. It is, in any case, clear that haemoglobin synthesis may be both stimulated and *suppressed* by the cytoplasm of the recipient cell, and the evidence, so far, appears to indicate that this regulation does not operate through the erythrocyte nucleus."*†
And earlier (p. 53) he also remarks that, in heterokaryons, combining mouse fibroblasts and mature erythrocytes "the cytoplasm of the recipient cell does not induce the erythrocyte nucleus to synthesize templates for haemoglobin synthesis; or, if it does induce the synthesis of these templates, it does not permit their translation into protein. This is hardly surprising. For when the recipient cell is a normal diploid cell, for example, a chick fibroblast, we must assume that its nucleus already contains the genes that determine the synthesis of haemoglobin; and since the fibroblast does not synthesize haemoglobin despite the fact that its nucleus contains the necessary genes, it is difficult to see why haemoglobin synthesis should be induced in the fibroblast cytoplasm simply by the introduction of a second set of the same genes."

The last quoted sentence obviously ignores the possibility that the functional state of the erythrocyte genome was most probably different from that of the fibroblast. As I pointed out earlier (p. 55), it is precisely the mechanism which creates these differences that is at the heart of the problem of determination. I suppose Harris ascribes these differences entirely to the influence of the cytoplasm and regards them as immediately reversible (Note 23).

I shall only add now that the more or less rapid *suppression* of hemoglobin synthesis observed by Cook in all of the heterokaryons described above is entirely in line with the observations on the behavior of other luxury functions in hybrids capable of multiplication, made mostly in my laboratory, about which I shall speak next. (I shall then return to Cook's observations on hemoglobin synthesis and to their significance, as I see it.)

* All italics are mine.

† Concerning the initial *rise* of hemoglobin synthesis in these heterokaryons, Harris suggests: "It is possible that the initial stimulation is due to the fact that the general metabolic level in a multiplying tissue culture cell is much higher than that in a differentiating erythrocyte, so that, once templates for the synthesis of haemoglobin are available in the heterokaryon, translation of these templates proceeds at a faster rate. . . . It is also possible that synthesis of haemoglobin is limited by the availability of haem [Bruns and London, 1965[18]] and that the tissue culture cell, which synthesizes haem for the production of cytochrome B and certain enzymes, may make more haem available." (pp. 53–55 in Ref. 101)

CHAPTER 6

Application of Cell Hybridization to the Study of Differentiation:
II. The Fate of Luxury Functions in Somatic Hybrids

> *. . . in the face of formidable technical difficulties, the study of differentiation either from the genetic or from the biochemical point of view has not attained a state which would allow any detailed comparison of theory with experiment. . . . The remarkable advances achieved by the methodology of cell cultures encourage optimism. The greatest obstacle is the impossibility of performing genetic analysis without which there is no hope of ever dissecting out the mechanisms of differentiation.*
>
> (Jacques Monod and François Jacob. 1961, p. 400)

I shall consider now the fate of luxury functions in hybrids capable of continued multiplication. Since we are thinking of the emergence and maintenance of these functions in differentiated cells in terms of regulation of gene activities, it may be well to begin by recalling what information the studies of formal genetics (Chapter 3) have provided on the behavior of the numerous household (ubiquitous) enzymes in such multiplying hybrids.

In general, crosses between cells (of the same or different species), differing either in the *forms* of, or by the *presence* vs. *absence* of household enzymes, give the results expected on the assumption that the parental cells differed only in *structural* genes. The two rules which, in accordance with expectation, can be deduced from these observations are: (1) co-dominance of different alleles of homologous genes and (2) complementation of nonallelic deficiencies (Note 24). On the contrary, surprisingly little evidence has been obtained pointing to the intervention of *regulatory* genes. Look again at Table 3.1, in which I have summarized the results of human linkage studies by Bodmer's and

71

Ruddle's groups; you will notice that only 1 out of the 15 genes listed appears to be a regulatory one ("ES-2 regulator").* And then, this single exception concerns the activity of a nonubiquitous enzyme, Esterase 2, which, in the mouse, is restricted to a small number of organs;[117, 212] all of the other enzymes listed in Table 3.1 are household enzymes! Aside from that, somatic hybridization which in principle is the best available genetic tool for the detection of regulatory genes in mammalian cells, has, to my knowledge, provided to this day only a single example of *possible* mutations of a regulatory element concerned with the production of a *household* enzyme[152] (Note 25).

The earliest observations pointing to the "orthodox," joint expression of household functions in hybrid cells were the point of departure of our first studies on the behavior of luxury functions in somatic hybrids: in the belief that differentiation involves regulatory mechanisms, we looked forward to finding evidence of interactions between parental genomes not observed with household markers. Anticipating, I may say that the situation turned out generally to be very different indeed when we turned to luxury functions, and that thereafter one of the chief purposes of our experiments became to establish the empirical rules which govern interactions when differently specialized cells are fused to form hybrids.

Remarks about the material

Before I turn to these experiments, I must point out that, with two notable exceptions which I shall cite, all the hybrids I am going to describe call for the following comments:

First, they result from crosses of cells exhibiting one or more differentiated functions with cells which do not exhibit them. The latter are usually fibroblasts, but I do not imply that fibroblasts are undifferentiated or undetermined cells.

Second, the parental cells exhibiting the luxury functions, the fate of which we are going to study in the hybrids, are neoplastic cells of various types, chosen because of the remarkable stability

---

* There is a possibility that the apparent linkage between the structural genes of peptidase B and LDH-B is spurious owing to the presence of a regulatory gene linked to one of the above. This possibility is, however, considered improbable by Ruddle *et al.*[211] and Santachiara *et al.*[215]

of luxury functions in such cells. This, as should be clear from what I said earlier (p. 50), calls for some reservations in the interpretation of the results. It is something my co-workers and I take very seriously, and that is why we are trying at present to make similar crosses with normal diploid differentiated cells. Meanwhile, I can only remind you that Cook and Harris' study on hemoglobin synthesis involved normal diploid erythrocytes and red cell precursors (pp. 67ff.) and that, as I already have said, their results[101] are formally in line with most of our observations involving neoplastic differentiated cells.

Lastly, I must mention that almost all of the hybrids to be discussed now are the results of *inter*specific crosses (chosen because of the possibility of identifying in them, as to the species of origin, both the chromosomes and a number of household enzymes). This is disturbing at first sight, but we shall see that there is good evidence to show that intraspecific and interspecific crosses, involving the same functions, give similar results. To make things shorter, I may add that the "household enzymes" of the cell hybrids I shall talk about have been used to test the functional activity of both parental genomes, and always gave positive results.

Melanoma × *fibroblast hybrids*

A first study along the indicated lines was performed by Richard Davidson, Kohtaro Yamamoto, and myself.[47, 48] It involved spontaneous crosses between cells of a Syrian hamster *melanoma* (RPMI 3460-3) and of three kinds of mouse fibroblasts (Cl.1D, B-82 and N-2), all from permanent lines (derived from C3H mice with pigmented skin and eyes). The first hybrids were selected owing to the presence, in the parental cells, of selective markers (BUdR or 8-azaguanine resistances, i.e. TK or HGPRT deficiencies). Later, virus induced fusion was used also.

The melanoma cells synthesize melanin, a luxury item; the fibroblasts, of course, do not. Melanin synthesis requires, as you know, the presence of dihydroxyphenylalanine (or dopa) oxidase. Numerous hybrids have been isolated and, whatever the conditions of culture, none of them were pigmented. Karyological study of the various hybrids showed that they all began with

chromosomal constitutions very close to the expected ones*
(Figure 6.1 and Table 6.1), and that at the time of the enzymatic
determinations to be described, they had lost only 5 to 15% of the
chromosomes (Table 6.1 last column). It appeared unlikely
therefore that the absence of pigment is the result of the (apparently random) chromosome loss, i.e. of the loss of the hamster
structural genes.† The functional activity of both parental
genomes in these hybrids was demonstrated by the electrophoretic
study of their LDH and MDH (malate dehydrogenase) patterns.
A priori, the absence of pigment could also be due to the origin
of hybrids from fusion of the fibroblasts with amelanotic variants
present in most melanoma lines. However, cloning on a large
scale of the particular melanoma (3460-3) used in our experiments
revealed no amelanotic variants, and this, taken together with
the relatively high mating rate of the melanoma cells (1 out of
800 in our first experiments) made this hypothesis quite unlikely.
This conclusion was further strengthened by repeating the crosses
with the help of Sendaï virus: in these experiments, 1 melanoma
cell out of every 25 took part in the formation of a viable hybrid,
and all of the hybrids were again unpigmented.[41, 42] Lastly, the
high ploidy *per se* of the hybrids cannot be incriminated because
2s melanoma cells are pigmented. The only necessary missing
control in *this* series of experiments, is the hybrid resulting from
fusion of two pigmented melanoma cells which we have been
unable to obtain thus far. (See however, pp. 136-38.)

As I already have said, one of the critical factors involved in
melanin synthesis is the enzyme dopa-oxidase. The results of determinations of the activity of this enzyme in the melanoma cells,
in the fibroblasts with which they were crossed, and in one hybrid
from each cross are shown in Table 6.2. (In this experiment, all
determinations were made on the 5th day after inoculation of the
cultures, when the dopa-oxidase activity is at a maximum in the
melanoma.) It can be seen that the hybrid cells show as little
dopa-oxidase activity as the parental fibroblasts. In fact, if the
value of dopa autoxidation is subtracted, negative values are

* More exactly, like most hybrids, they contained when first examined (i.e.
after about 20 generations) slightly fewer chromosomes than expected.

† That this assumption was probably correct will be seen later, when we
examine the question of the reversibility of the changes induced by hybridization (pp. 89ff.).

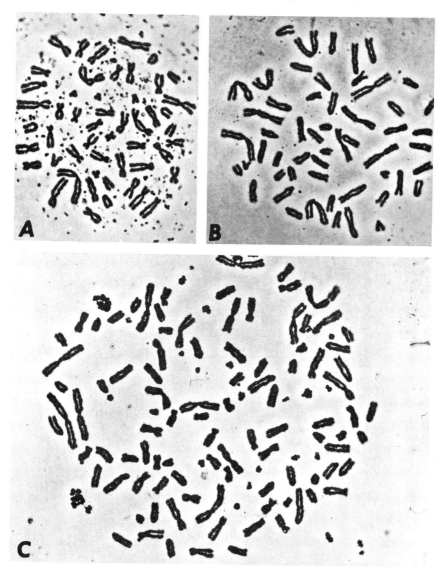

FIGURE 6.1. Metaphases of parents and hybrids of the melanoma $\times$ fibroblast cross. *A*, melanoma 3460-3; *B*, Cl.1D, *C*, hybrid. From Davidson *et al.*,[48] with permission of the Wistar Institute Press.

TABLE 6.1. Karyotypes of melanoma hybrids.[a]

Cell type	Initial number of chromosomes[b]						Loss[c] (%)
	Metacentric		Acrocentric		Total		
3460–3	43	(41–43)	8	(7–9)	51	(50–51)	
Cl.1D	9	(8–11)	43	(38–46)	52	(47–54)	
B-82	20	(19–22)	33	(30–36)	54	(51–59)	
N-2	20	(18–21)	33	(30–37)	52	(50–55)	
Hybrid clones							
3460–3/Cl.1D	52	(50–55)	48	(43–55)	101	(96–105)	5–8
3460–3/B-82	65	(59–68)	35	(26–39)	99	(92–102)	6
3460–3/N-2–3	58	(54–61)	34	(32–40)	94	(86–100)	10–15

[a] From data of Davidson *et al.*[48]

[b] Modal number and range (in parenthesis).

[c] At time of enzyme determinations. Where a range of values is given, different values relate to variations between clones or to different numbers of cell generations since their isolation. The percent of chromosomes lost is calculated with respect to *expected* total chromosome numbers (i.e. on the assumption that, *ab initio*, the total chromosome number in the hybrids equaled the sum of the *modal* chromosome numbers of the parental lines).

TABLE 6.2. Dopa-oxidase activity of melanoma hybrids[a]

Cell type	Activity[b]
3460–3	169
Cl.1D	4
B-82	5
N-2	4
Hybrids	
3460–3/Cl.1D	5
3460–3/B-82	4
3460–3/N-2	3
Dopa autoxidation	11

[a] Recalculated from data of Davidson *et al.*[47]

[b] Expressed as change in optical density/500μg protein/hr.

obtained. This has been found to be due to the presence of an inhibitor. However, it was shown by Davidson and Yamamoto in a series of experiments on the kinetics of dopa oxidation in mixtures of extracts of melanoma cells with extracts of Clone 1D or of hybrid cells, that this inhibitor is not responsible for the absence of dopa-oxidase activity in the hybrids; moreover, the results of these experiments suggest that the absence of dopa-oxidase activity in the hybrids is not due to interactions at the level of enzyme activity, enzyme activation, or stabilization; and it appears probable, but by no means demonstrated, that it is due to the absence of dopa-oxidase synthesis.[49]

Whether the absence of dopa-oxidase is the only block to pigment production in the hybrids remains unknown. Be this as it may, it appears that hybridization with cells not producing pigment results in the *extinction* of this luxury function of (neoplastic) melanocytes. That this phenomenon has nothing to do with the interspecific nature of the crosses is clear from the fact that Silagi[225] has obtained exactly similar results with hybrids between a *mouse* melanoma (B 16) and (HGPRT deficient) *mouse* fibroblasts of Littlefield's line A-9. The chromosome numbers in these hybrids were very near to the expected, and the functioning of the melanoma genome demonstrated by the production of HGPRT and of histocompatibility ($H-2^b$) antigens characteristic of the mouse line (C57BL) in which the melanoma originated.[225]

In our very first publications (1966) on the melanoma hybrids, we were "inclined to conclude that a step in the process of pigment formation in the cells studied is under negative control,"[*] and raised the following questions:

" (a) Is the repression[†] of the pigment-forming process in the hybrids dependent upon the continued presence of certain genes of the unpigmented mouse cells? (b) At what point in this process is the negative control exerted? (c) Are other processes associated with the differentiated state of pigment-producing cells affected by hybridization in a similar manner? (d) Can our conclusion be extrapolated to the regulation of pigment formation in normal melanocytes? (e) Can it be extended to other types of differentiation."[47][§]

In retrospect, I note with some satisfaction that several of these questions have been answered by now by experiments to be described below.

The notion of extinction of luxury functions

Formally, production of pigment and dopa-oxidase activity behave, in the hybrids with nonpigmented cells, as *recessive* traits. I speak of "extinction" of the functions rather than of their "recessiveness," because the nature of the changes underlying

---

[*] I shall return below (pp. 96f.) to the validity of this conclusion.

[†] Referred to in these lectures as "extinction."

[§] We also referred in this paper to earlier work on collagen synthesis which will be discussed later (p. 141).

determination and differentiation remain unknown and is pre-cisely what we want to find out, and because, as I said earlier (pp. 53-54), it is unlikely that they result from gene mutations to which the terms "dominant" and "recessive" are applied. On the other hand, I prefer to use the term "extinction" rather than "repression" because the latter refers to a precise mechanism of regulation (at the transcriptional level)[119] so far demonstrated only in bacteria and their viruses.

I think this new notion of extinction is important because it might eventually serve as a "handle" for the biochemical defini-tion of the phenomenon described by this term and, hence, of the molecular mechanisms operating in differentiation.

Other instances of extinction of single luxury functions

Total extinction of a luxury function was observed also by Sonnenschein and co-workers[237] who crossed mouse fibroblasts (Clone 1D) with cells of a *rat pituitary tumor*, adapted to growth *in vitro* (line GH_1 $2C_1$).[245] These (hypotetraploid) tumor cells synthesize and secrete into the medium a (protein) growth hor-mone which can be measured immunologically. A hybrid cell line (a clone or a mixture of clones?—compare Refs. 235 and 237) was obtained whose karyotype comprised almost exactly two Cl.1D and one GH_1 $2C_1$ genomes. No hormone secretion by the hybrid cells was detected.

A less clearcut case of extinction of a luxury function was ob-served in hybrids between mouse fibroblasts of line Cl.1D and *glial* cells of a line isolated from a chemically induced rat brain tumor.[9] Such hybrids have been recently produced and studied in my laboratory by Benda and Davidson,[8] first with respect to protein S-100,* a soluble, highly acidic protein of molecular weight 24,000[175] which has been shown to be present in the brains of a wide variety of (and probably all) vertebrates,[148] where it appears to be restricted to glial cells.[118, 196] It is appropriate to add that a study of purified beef S-100, has suggested that it is made up of several components.[24]

In cell extracts, S-100 can be detected and measured by a technique of complement (C') fixation, using antiserum prepared

---

* Experiments by Davidson and Benda on two other luxury functions of glial cells will be mentioned on pp. 83 and 135-136.

against purified beef S-100. The relative amounts of S-100 in extracts of different cells can be deduced from the concentration of soluble protein required for maximum fixation of complement (which is considered as the point of antigen-antibody equivalence), provided the *amounts* of complement fixed at this point be the same.

When cell lines other than RG6A (the clone of glial cells used in these experiments) were tested for S-100, the concentration of soluble protein required for maximum C' fixation (Q_{max}) was compared to the concentration of soluble protein of RG6A (R_{max}) which gave maximum C' fixation. The serological activity of the line tested is defined as (R_{max}/Q_{max}) × 100. Thus, by definition, the serological activity of RG6A is equal to 100. If an extract produces maximum C' fixation at the same concentration of protein as RG6A, its serological activity is 100, and if twenty times more protein is necessary to obtain maximum C' fixation with a given extract than with an extract of RG6A, its serological activity is 5.

As can be seen in Table 6.3 (column "Serological Activity"), glial cells (RG6A) grown *in vitro*, produce large amounts of S-100; fibroblasts of line Cl.1D, surprisingly, appear to contain 3% of the amount present in glial cells, and hybrids between them

TABLE 6.3. Serological activities in glial cell hybrids[a]

Cell type	Serological activity[b]	Protein (μg)/cell	Activity/cell[c]
RG6A	100	1.1×10^{-4}	100
RG6², Cl.1	65	2.2×10^{-4}	130
Cl.2	135	2.2×10^{-4}	270
Cl.1D	5	1.3×10^{-4}	6
Hybrids			
RG6 × Cl.1D	8	2.4×10^{-4}	17
RG6²-2 × Cl.1D	9	3.4×10^{-4}	28

[a] Data of Benda and Davidson.[8]

[b] Averages for several clones of a given class except for RG6². Since the two RG6² clones have very different values, the individual figures are given.

[c] Activity/cell is the serological activity multiplied by the protein/cell, and expressed relative to the value in the RG6A cells, which is set equal to 100.

5 to 10%. Thus, at first sight, we have here a case of only *partial* extinction of a differentiated function, the production of S-100. However, the significance of the measurements of serological activities of the fibroblasts and of the hybrids is questionable

79

because the amounts of complement fixed by extracts of these cells at the point of maximum fixation are distinctly lower than those fixed by extracts of glial cells. This is shown in Figures 6.2 and 6.3. The first of these figures shows the results of C' fixation

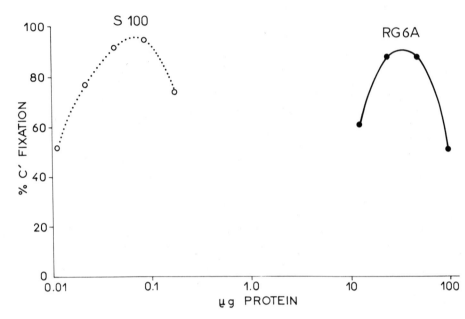

FIGURE 6.2. Complement fixation curves obtained with purified beef S-100 and with an extract of glial cells RG6A. Redrawn from Benda and Davidson[8] with permission of the Wistar Institute Press.

by an extract of glial cells (RG6A) and by purified S-100. It can be seen that glial cell extract fixes almost as much C' as purified S-100. Figure 6.3, on the contrary, shows that, by comparison with that of glial cells, maximum C' fixation is clearly reduced in the hybrids and is very small in Clone 1D fibroblasts. This makes it impossible at the moment to say whether the complement fixing material in the fibroblast and/or hybrids is some cross-reacting material unrelated to S-100 or represents S-100 protein more or less modified in its conformation and immunological behavior (such changes have indeed been observed to occur in purified S-100 as a result of heat denaturation at 50–60° C[135]). The solution of this question will require at least partial purification of the complement fixing material of the hybrids, which is

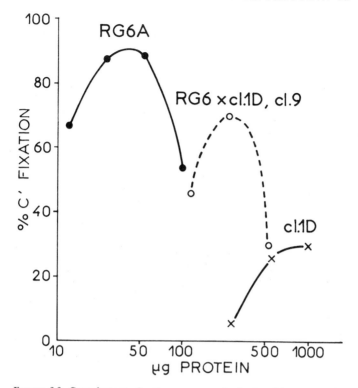

FIGURE 6.3. Complement fixation curves obtained with extracts of: glial cells (●), Cl.1D fibroblasts (×), and hybrid cells (O). From Benda and Davidson,[8] with permission of the Wistar Institute Press.

being undertaken. Meanwhile, however, it is impossible to say whether we are dealing here with total extinction which is only spuriously partial or, at the other extreme, with production of a full amount of S-100, modified in its properties* (Note 26).

Needless to say, in this study also it has been ascertained that what formally appears as partial extinction of S-100 is due neither to the cell fusion itself, nor to the increased ploidy of the cells. This is shown by the high serological activity of 2s glial cells RG6A[2], obtained by virus induced pairwise fusion of 1s glial cells (Table 6.3). Lastly, it is unlikely that the (partial?) extinction of S-100 is due to chromosome loss. Table 6.4 shows that the RG6A cells have a modal chromosome number (42) which corresponds to that of diploid rat cells. The 2s glial cells (RG6[2]) contain ap-

* This modification may involve either the whole molecule of S-100 or only one of its subunits.

81

TABLE 6.4. Karyological characteristics of glial cell hybrids[a, b]

Cell type	Telo-centric	Large meta-centric	Large subtelo-centric	Small meta-centric	Small subtelo-centric	Total
RG6A	18	0	4	14	6	42 (41–42)
RG6[2]	34	0	8	25	12	79 (76–82)
Cl.1D	44	9	0	0	0	53 (51–53)
Hybrids						
RG6 × Cl.1D	60	9	4	14	6	92 (85–95)
RG6[2] × Cl.1D	60	12	8	24	9	114 (105–122)

[a] Data of Benda and Davidson.[8]

[b] The values represent the modes for all the clones together of a given line or hybrids. The numbers in parentheses following the total chromosome number indicate the range within which 90% of the cells fall. Because the numbers of chromosomes of each type for the RG6[2] × Cl.1D hybrids are distributed throughout the range without forming a distinct mode, the values given for these hybrids and for the RG6[2] parent are means and not modes.

proximately twice as many chromosomes of each type as the $1s$ cells. The number and distribution of chromosomes in the hybrids is very near the expected one: there is very little chromosome loss (in particular of "rat markers"). The loss of chromosomes in the hybrids between $2s$ glial cells (RG6[2]) and Cl.1D fibroblasts is of the order of 15% and is chiefly at the expense of telocentric chromosomes, most of which are contributed by the mouse parent.

Another case of extinction of a luxury function has been described by Klebe *et al.*[136] The function in question is the production of an esterase (ES-2) which is restricted to a small number of organs of the mouse, among them, the kidney. Klebe *et al.* have studied ES-2 in hybrids between cells of a mouse kidney (adenocarcinoma) line which exhibit this function very actively and either mouse or human fibroblasts, which do not. I shall defer the presentation of this case because it is complicated (in a most interesting way!) by rapid chromosome losses in some of these hybrids (see below, pp. 93f.).

Fate of multiple liver functions in
hepatoma × fibroblast hybrids

I have thus far presented evidence on the fate in hybrids of *one* differentiated function characteristic of each cell type. But, as often emphasized by Grobstein,[85] a differentiated cell of a given type is characterized in reality by a whole set of properties which make its "enzymatic profile" different from that of any other cell type. It would be of interest therefore to establish whether the extinction of specialized functions, of which I have just described a few examples, is observed simultaneously for all differentiated functions of a given "set" (and hence, assuming, as I do, that it is a regulatory phenomenon, to establish whether multiple functions are subject to coordinate control).

This question is being investigated in my laboratory by Drs. P. Benda and R. L. Davidson and by a team headed by Mary Weiss on hybrids of glial and hepatoma cells, respectively. Since in the remaining time I cannot describe the results of both these studies, I shall speak only of the work on the hepatoma hybrids which by now has progressed much further. (For the work on glial cells, see Note 27.)

For the study of hepatoma hybrids, a cell line was chosen derived from a minimum deviation rat *hepatoma* (Reuber H-35) adapted to growth *in vitro* by Pitot and co-workers.[199] Cells of this line exhibit a number of luxury functions typical of normal ·liver parenchyma cells. Among these, three have been investigated so far:

1. *Production and inducibility of tyrosine aminotransferase* or TAT. This enzyme, although not strictly limited to liver, presents in this organ a particularly elevated activity, which (in liver *only*) can be further increased (or induced) by glucocorticosteroid hormones.[149] The work of Tomkins' and Kenny's groups has shown that this induction represents an actual increase of TAT synthesis (for references, see review by Tomkins *et al.*[247]). As you will see, this enzyme is also produced in large amounts by, and is inducible in, the hepatoma cells used in the experiments I shall describe.

2. *Production of fructose-1,6-diphosphate aldolase.* This is a tetrameric enzyme[190] present in different forms in different organs.[29, 191] The different forms, identifiable by electrophoresis, result from combinations of three types of monomers, A, B, and C, specified by as many nonallelic structural genes. The C form being confined to the brain and to certain embryonic tissues, we are going to be concerned only with A and B. Of these two, A is

ubiquitous; B, although present also in the kidney and the intestine, is particularly high in the liver. In the hepatoma cells we are concerned with, aldolase A activity is higher than in the liver; but the presence of appreciable amounts of aldolase B is easily detected electrophoretically by the occurrence of AB heteropolymers. Moreover, on the basis of spectrophotometric assays, calculations of the relative proportions of A and B can be made owing to the different affinities of aldolases A and B for their two substrates, fructose-1,6-diphosphate and fructose-1-phosphate (FDP/F1P ratio).[191]

3. *Production of serum albumin.* This is probably the most specific of all liver functions (review in Ref. 193); albumin is secreted by hepatoma cells into the medium where its amount can easily be measured by immunological techniques.[39, 249]

The first experiments involved crosses between cells of a hepatoma clone, designated Fu5, and the thymidine kinase deficient mouse fibroblasts of the permanent line 3T3-4E.[162] Numerous hybrid clones have been obtained by the use of Sendaï virus and of the half selective system (Chapter 2, p. 13).

Since the hybrid clones generally present similar properties, I am going to speak in most cases about all of them together, except when differences between clones are significant.

All of these hybrids contain somewhat fewer chromosomes than expected and, since about half of the rat chromosomes are distinguishable from those of the mouse parent, it appears probable that they all have lost mostly rat chromosomes (up to 20% of the latter).

Let us see now what becomes of the liver luxury functions in the hybrids.

The results of determination of *TAT activity*[220] of extracts of parental and hybrid cells in late log phase, either after growth in normal medium or after exposure for 24 hours to $10^{-6} M$ Dexamethasone, are given in Table 6.5. You can see that the TAT specific activity and activity per cell (expressed in terms of $m\mu$-moles of p-hydroxyphenylpyruvate formed) is high in uninduced hepatoma cells; that in the 3T3 fibroblasts, this baseline activity is low; and finally that in the hybrids, the baseline is about twice that of the fibroblasts. I may add that the heat inactivation curves of rat and mouse TAT are different, and it has thus been possible

84

Table 6.5. Tyrosine aminotransferase in hepatoma hybrids[a]

Cell type	mU[b]/mg protein		mU[b]/10^6 cells	
	Baseline	$10^{-6}\,M$ Dex	Baseline	$10^{-6}\,M$ Dex
3T3	0.8	0.77	0.2	0.19
Fu5	24.7	154	3.9	30
Hybrids				
3F11	1.6	1.5	0.5	0.5
3F14	1.8	1.5	0.7	0.5
3F16	1.0	0.7	0.5	0.3
Averages of 8 clones	1.6 (0.9–1.8)	1.6 (0.7–1.9)	0.64 (0.5–0.8)	0.55 (0.3–1.1)

[a] Data of Schneider and Weiss.[220]
[b] mU = mμmoles of *p*-hydroxyphenylpyruvate/min at 37°.

to establish that the baseline activity in the hybrids is due to the presence of both rat and mouse enzymes.

From the data of Table 6.5, you can see also that, in the presence of Dexamethasone, the TAT specific activity of the hepatoma cells increases by a factor of 6, while no induction is detected either in the fibroblasts or in the hybrids; since the degree of induction depends on the growth phase of the culture, I hasten to add that my statement about the absence of TAT induction in the hybrids holds for determinations made at *any* point of the growth cycle. Lastly, I want to point out that the presence of rat TAT in the hybrids, which are not inducible, shows that baseline and inducibility are subject to independent controls. In fact, we shall soon see that these two characteristics segregate independently: they must therefore be controlled by two different genes. This is a conclusion earlier arrived at by Tomkins as a result of his studies of TAT in hepatoma cells.[247] Moreover, the model proposed by Tomkins postulates the production, by one of these genes, of a post-transcriptional repressor which, in the absence of steroids, inhibits the translation and promotes the degradation of TAT messenger RNA.

The data concerning *aldolase*[10] are given in Table 6.6. They show that the hepatoma cells (Fu5) have considerable aldolase activity and that 1% of this activity is due to aldolase B. The

TABLE 6.6 Fructose-1,6-diphosphate aldolase in hepatoma hybrids[a]

Cell type	Specific activity[b]	Calculated % aldolase B[c]
Fu5	103	1.0
3T3	145	<0.02
Hybrids[d]	150 (120–172)	<0.025

[a] Compiled from data of Bertolotti and Weiss.[10]

[b] $m\mu$moles of FDP cleaved/min/mg protein at 26°.

[c] Based, in the hepatoma cells, upon the substrate activity ratio (FDP/F1P). In the other two cases, the values given are based on the minimal activity detectable by electrophoresis in mixed extracts.

[d] The value of specific activity given here is an average of that of 8 clones; the range of values is in parentheses.

3T3 fibroblasts have a total aldolase activity even higher than that of the hepatoma cells, but no detectable amounts of aldolase B; and the hybrids present exactly the same picture as the fibroblasts. I may add, on the one hand, that the method used is sensitive enough to reveal aldolase B activity less than 10% of that of the hepatoma, and, on the other, that the aldolase A activity of the hybrids has been shown by heat inactivation experiments to be due to both rat and mouse enzymes.

Finally, some of the data concerning *albumin* secretion[195] are given in Table 6.7. They show that hepatoma cells secrete considerable amounts of rat albumin; that 3T3 fibroblasts secrete none (and separate tests show that they do not secrete mouse

TABLE 6.7. Rat albumin synthesis in hepatoma hybrids[a]: analysis by micro-complement fixation[b]

Cell type	Albumin produced[c] (μg/72hr/10^6 cells)
Fu5	4 (3.75–4.7)
3T3	<0.003
Hybrid clones	
3F11	1.5 (1.4–1.6)
3F14	0.6 (0.5–0.7)
3F16	0.36 (0.32–0.41)

[a] Data of Peterson and Weiss.[195]

[b] Anti-serum: Rabbit anti-rat albumin.

[c] All determinations performed in 3 independent experiments.

albumin either); and that different hybrid clones secrete rat albumin in variable amounts which, depending on the particular clone, may represent from 5 to 30% of the albumin secreted by the hepatoma. None of these clones produce mouse albumin.

Summing up, we may then say that, just like in the cases I described earlier, there is total or partial extinction of the specialized liver functions investigated so far in the hybrids between hepatoma cells and mouse fibroblasts (high level of TAT, TAT inducibility, synthesis of aldolase B and albumin). However, I must point out that this extinction is *very* incomplete in the case of albumin, the most specific of all liver functions, since in some clones its production still represents *ca.* 30% of that of the hepatoma. This makes one wonder whether we have here an indication of independent regulation of TAT (baseline and inducibility) and aldolase B, on the one hand, and of albumin synthesis on the other, or whether the latter is subject to an altogether different mechanism of regulation. (Further evidence indicating relative independence of the controls of the different liver specific functions under consideration will be given on p. 95.)

That total or partial extinction of differentiated functions is not due to chromosome losses has already been shown for the melanoma and glial cell hybrids. In the former case, it was also shown that extinction is not due to the fact that we are dealing with *inter*specific hybrids (see p. 77). Further evidence to the same effect is provided by as yet unpublished observations on an *intra*specific hepatoma hybrid and its derivatives.[255] I shall describe these observations in some detail because they are interesting also in another respect.

Intraspecific hepatoma hybrids

The hybrid which is at the origin of the studies to be described was obtained from a cross of Fu5-5 (a subclone of hepatoma Fu5) with cells of an epithelial cell line isolated a few years ago by Coon from the liver of a young Buffalo rat. This line was assumed to have originated from liver parenchyma (and consequently designated as BRL: Buffalo rat liver). It is diploid and is said to have displayed, to begin with, some liver specific functions.[33, 34] By now however, it synthesizes none of the liver specific products for which it has been tested (in particular, none of those produced by the hepatoma cells with which it has been crossed). Thus, BRL may be in reality either a line of "dedifferentiated" liver parenchyma cells, or of some other epithelial cells of liver origin (which represent about 30% of the liver). Be this as it may, it is a *rat* line, and it is *diploid*, as witnessed by its karyotype.

Although the chromosomes of Fu5-5 and BRL cannot be distinguished, it is clear that the hybrids contain the chromosome complements of both parents: the total number of chromosomes is remarkably near the expected sum of parental numbers. This can be seen in Table 6.8 which shows, moreover, that under the usual

TABLE 6.8. Hepatoma hybrids: numbers of chromosomes in parental and hybrid cells[a]

Cell type	Large submeta-centric	Satellited	Large meta-centric	Small metacentric	Telocentric	Total
BRL-1	2	2	0	20	18	42
Fu5-5	2	2 (1–3)	1	23 (21-24)	24 (22–25)	52 (51–53)
Hybrid						
Expected	4	4 (3–5)	1	43 (41–44)	42 (40–43)	94 (93–95)
Observed in clone BF5 after [b]:						
1 month (22)	4	4.3 (4–5)	1	41.8 (40–43)	40.8 (40–42)	92 (91–93)
2 months (35)	4	4	1	41.8 (40–43)	40.8 (40–42)	92 (91–93)
4 months (75)	4	3.6 (3–4)	1	38.6 (34–43)	44.2 (41–48)	91.4 (89–93)

[a] Compiled from data of Weiss and Chaplain.[255]
[b] Numbers of cell generations in parentheses.

culture conditions, the karyotype of the hybrid cells is extremely stable—there is no chromosome loss on prolonged cultivation.[255]

Among the hepatoma functions studied in the previous series of experiments, only TAT and aldolase have been tested so far in some detail in the Fu5-5 × BRL hybrids.[255] In these hybrids, like in those with 3T3, the baseline of TAT activity is very low, and TAT is not inducible: I shall give you the data in a minute. Aldolase B is also absent, while some albumin is produced (J. Peterson, personal communication). Thus, the properties of *intra*specific hepatoma hybrids with a highly stable karyotype appear to be exactly similar to those of *inter*specific hepatoma hybrids. Moreover, the similarity of the properties of the two types of hybrids suggests that the mechanism which, in the hybrids, prevents the full expression of the genes involved in the differentiated hepatoma functions, is similar in 3T3 fibroblasts and in the epithelial BRL cells[255] (Note 28).

Serependity

In one respect I have not mentioned yet, the Fu5-5 $\times$ BRL hybrids are different from the Fu5 $\times$ 3T3 hybrids: contrary to the latter, they *are* contact-inhibited. Now, in an attempt to isolate variant hybrids lacking contact inhibition (which, from earlier observations were known to be characterized by the loss of many chromosomes[260]), Mary Weiss[255] last year kept renewing, for two to three consecutive months, the medium of some cultures of Fu5-5 $\times$ BRL hybrids, which had reached saturation density. She did, in this way, isolate 14 such variant hybrids, but, in addition, discovered a most unexpected and interesting correlation between karyotype and TAT inducibility.

The first column of Table 6.9 shows the chromosome numbers of the parental cells and of the hybrids. Notice that the original hybrid BF5 and a subclone (BF5-C) derived from it have almost exactly the expected total number of chromosomes. The two last hybrid clones (BF5-β and BF5-1-1) are representative of the 14

TABLE 6.9. Tyrosine aminotranferase in hepatoma hybrids[a]

Cell type	No. of chromosomes	mU[b]/mg protein	
		Baseline	Induced[c]
Fu5-5	52 (51–53)	22.0	318.0 (14X)
BRL-1	42	0.27	0.22
Hybrids			
BF5	92 (91–93)	0.78	1.06
BF5-C	92 (91–93)	0.58	0.57
BF5-β	59 (58–59)	0.75	0.8
BF5-1-1	63 (60–66)	0.5	6.0 (12X)

[a] Compiled from data of Weiss and Chaplain.[255]
[b] mU = mμmoles of *p*-hydroxyphenylpyruvate/min at 37°.
[c] 10^{-6} *M* Dexamethasone.

"segregated" populations of BF5 obtained in the way I just told you. You can see that they have lost about 30% of the initial number of chromosomes. Whether they lost hepatoma or BRL chromosomes is unknown, of course, because both parental cells are rat cells carrying no special markers. It is, however, highly likely that at least some of the chromosomes lost were of BRL origin since the growth habit and the degree of malignancy of some of the segregated hybrids are very similar to those of the hepatoma parent.

Look now at the TAT activity of the different hybrids, given in the two right hand columns of Table 6.9. You can see, first, that the parental hepatoma clone used in these experiments has a rather high baseline activity (22.0) and that this activity is increased about 14 fold by exposure to Dexamethasone. You can also see that, as I said earlier, BRL has a low baseline and is not inducible. The same can be said about the two complete hybrids (BF5 and BF5-C) and the first segregated hybrid (BF5-β). Twelve other segregated hybrids behaved like BF5-β. However, as the last line of the table shows, the fourteenth (BF5-1-1), although it also has a very low TAT baseline as compared to the hepatoma, *is* inducible: in the presence of Dex, its TAT activity increases about 12 fold with respect to the baseline, like that of the parental hepatoma!

A study of TAT inducibility in the hepatoma and in the "segregated" hybrid cells has shown that both cell types undergo, in this respect, significant variations in the course of the growth cycle. Figure 6.4A shows that in the hepatoma cultures, maximum inducibility is observed to occur during early logarithmic growth; however, both in this phase and in the stationary* one, maximum induced TAT activity is observed after 20–30 hours of exposure to Dex; one will notice also that once the maximum is reached, the level of activity remains high. By contrast, hybrid cells (BF5-1-1) in the stationary phase show an extremely rapid increase in TAT activity and a precipitous drop after *ca.* 8 hours in the presence of inducer (Figure 6.4B). However, this peculiar phenomenon is unique to hybrid cells in the stationary phase,

---

* Both the hepatoma and the hybrid cells lack contact inhibition; cultures of these cells are therefore never really "stationary" if the medium is renewed, but their multiplication becomes very slow in crowded cultures.

90

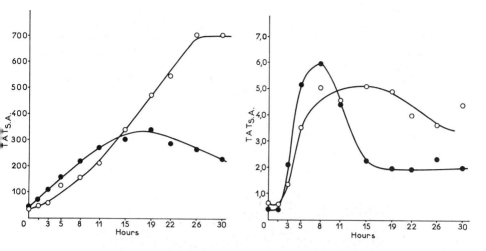

FIGURE 6.4. Time course of TAT induction in hepatoma cells (left) and in a "re-expressing" hybrid (right). For both cell types, two curves are shown, representing early logarithmic cultures (*open circles*) and late logarithmic or stationary cultures (*closed circles*). From Weiss and Chaplain,[255] with permission.

and is not observed during their logarithmic growth. Comparison of the curves in Figure 6.4A and B will show that early log hybrid and stationary hepatoma cells exhibit rather similar kinetics of induction. (A new hybrid subclone, recently derived from that used in the experiments illustrated in Figure 6.2B, no longer shows the peculiar kinetics of induction in the stationary phase: the TAT activity remains high even after 20–30 hours in Dex, like in the log phase cells. It is interesting to note that this new hybrid has lost more chromosomes.[255]) Additional experiments have shown that the induction of TAT in parental (Fu5-5) and hybrid (BF5-1-1) cells is similar in all of the following respects: inhibition by Cycloheximide, by Actinomycin D, and response to varying concentration of Dexamethasone.[255] *

Recent experiments by Bertolotti and Weiss (personal communication) have shown that re-expression of aldolase B similarly occurs in some of the segregated Fu5-5 × BRL hybrids; that its re-expression is in some cases independent of that of TAT inducibility; and further that

* However, although the intracellular distribution and heat stability of TAT from induced or noninduced hepatoma cells, and from induced hybrid cells are similar, TAT from uninduced hybrid cells shows different properties, rather similar to those of TAT from BRL.

91

the activity of aldolase B in these hybrid cells is as high or higher than that of the parental hepatoma cells. Like the hybrid clones which show TAT inducibility, those which express aldolase B have undergone substantial reductions in chromosome number and retain only 52 chromosomes.

Figure 6.5 shows aldolase zymograms of the parental cells (Fu5-5 and BRL-1), the chromosomally complete hybrid BF5, and a segregated

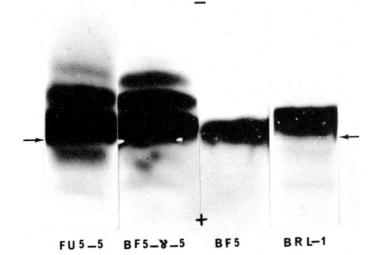

FU 5 —5   B F5_γ_5   B F5   B R L—1

FIGURE 6.5. Re-expression of aldolase B. Explanation in text. Courtesy of R. Bertolotti and M. C. Weiss.

hybrid (BF5-γ-5) which produces aldolase B and shows no TAT inducibility. It can be seen that BRL-1 and BF5 show only a single band of activity, corresponding to pure aldolase A, and that Fu5-5 and BF5-γ-5 show, in addition, two of the AB heterotetramers.

These observations clearly indicate a great deal of independence of the controls of TAT inducibility and of aldolase B production, as has already been shown for albumin.* (Note that this is reminiscent of the independent variation of characters in tumor progression emphasized by Foulds.[74])

* Where there is independent re-expression of TAT inducibility and of aldolase B production in the segregated BF hybrids it could be due either to independent loss of the relevant structural genes or to their independent "switching on." In these cases, the intraspecific nature of the hybrids precludes a distinction between these two possible mechanisms. However, the second mechanism is certainly involved, in the case of one segregated clone which, *in vitro*, shows TAT inducibility in the absence of detectable aldolase B. Inoculated into rats, this clone formed tumors in which aldolase B is re-expressed showing that its absence *in vitro* was not due to the loss of the relevant structural gene. This clearly points to the existence of separate switches for different luxury functions of a given "set." This does not imply of course that there do

The important conclusion suggested by these observations is that *the mechanism which results in the extinction of a given tissue specific function requires the continued presence of some BRL chromosomes.* Since the chromosomes of the two parents of these hepatoma hybrids cannot be distinguished, it is impossible in this case to establish an exact correlation between karyotype and phenotype. However, the same conclusion has been arrived at recently by Ruddle's group as a result of a study of hybrids (WR) between cells of a mouse *renal adenocarcinoma*, adapted to growth *in vitro* (line RAG, resistant to 8-azaguanine),[137] and human diploid fibroblasts WI-38.[136] *

The identification of hybrids was based on the following considerations. The hybrid must retain the human X-chromosome because it carries the human HGPRT+ gene necessary for their growth in HAT. Since the human X also carries the G6PD gene (Table 3.1) which specifies a form of glucose-6-phosphate dehydrogenase electrophoretically distinguishable from that of the mouse, all hybrid clones were identified by (1) the presence of both mouse and human G6PD (and hybrid enzyme molecules) and (2) a complement of human biarmed chromosomes which included a recognizable X (a single exception was one clone, WR-5, which appears to have a deletion of one of the arms of the mouse X-chromosome; cf. Ref. 136).

---

not exist superimposed general switches for the ensemble of functions of the "set," i.e. that normal development of a given cell type may involve the intervention of a hierarchy of switches. In fact, as Markert[160] suggested in the following quotation, the existence of a general switch in the programmed development of higher organisms can be deduced from the nature of teratomas:

"[These] tumors are chaotic arrangements of many different cell types, frequently organized as functioning tissues. . . . Clearly, the neoplastic event in teratoma formation occurs at a very early stage in cell differentiation—in the primordial germ cell. The occurrence of a neoplastic event at this early stage, before the inception of any embryonic development, suggests the existence of a single basic common mechanism for programming genes. A defect in this postulated mechanism would produce the chaotic arrangement of tissues seen in teratomas but would not prevent normal differentiation of specific cell types once the initial defect had been passed. That is, specific programs for regulating gene function in highly differentiated cells can still develop in teratomas but the selection of a particular program appears uncontrolled and erratic."[160]

* The observations of Klebe *et al.*[136] on the re-expression of ES-2 had been published prior to my lectures, while the data on re-expression of TAT inducibility[255] and aldolase B had not. I talked first about the latter data in order not to disrupt the presentation of the various results obtained with the hepatoma hybrids.

RAG cells, like some cells in one part of the normal kidney, synthesize an eserine-insensitive esterase, ES-2, which is restricted to a few organs (kidney, intestine, liver), and which can be identified electrophoretically.[117, 213] ES-2 is not detectable in the human fibroblasts. When first examined about two months after their isolation, several independent hybrid clones exhibited the ES-2 band with its characteristic conformer, and only one did not. This is shown in Figure 6.6A. You can see that the ES-2 band is

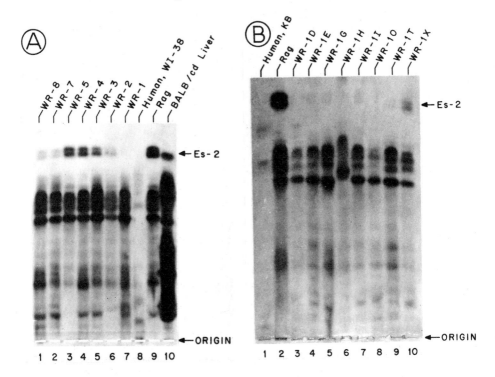

FIGURE 6.6. Re-expression of ES-2. Explanation in text. From Klebe et al.,[186] with permission.

present in RAG (and shows the conformer which is not present in mouse liver); that it is absent in the human cells WI-38; and, lastly, that it is present in all hybrid WR clones except WR-1.

Since human × mouse hybrids lose human chromosomes, it was thought that WR-1 may be the only clone which has retained the human chromosome causing the extinction of ES-2. Therefore, three hybrid clones exhibiting ES-2 were subcloned, as well as clone WR-1 which did not. Twenty-two subclones of the former three clones were again found to be positive (ES-2+). The sub-clones of WR-1 however gave a different picture: as you can see in Figure 6.6B, in most of the subclones ES-2 activity was absent (ES-2−) or faintly visible; but in one of them (WR-IX, on the extreme right), the ES-2 band was fully re-expressed and exhibited the characteristic conformer.

As a control, a cross was made between RAG and Clone 1D (which has very low ES-2 activity :<12% of that of RAG). The hybrid (RELM) obtained, exhibited ES-2 activity no higher than that of Clone 1D (<10% of RAG).

Detailed karyotypic analysis of two ES-2− clones and of six ES-2+ clones from the RAG × WI-38 cross has been made and led the authors to the conclusion that the extinction of ES-2 in the hybrids is most probably correlated with the retention of human chromosome C_{10}.* It is no doubt significant that in this work too it was noticed that the cell morphology and growth habit of the hybrids with low numbers of human chromosomes are similar to those of the mouse parent (Note 29).

Significance of the re-expression of luxury functions

The re-expression, upon loss of (probably definite) chromosomes, of luxury functions (TAT inducibility and ES-2) which were "extinguished" in these hybrids, proves that extinction is not due to loss of the structural genes coding for the corresponding proteins; furthermore, and more importantly, it shows the amazing stability of what I have called the *epigenetic states* which are the results of determination. The different epigenetic states must be endowed with a high degree of stability since, in the hybrids, they are apparently propagated through very many cell generations *without being expressed*. Lastly, taken together with the fact that the continuous presence of certain, apparently specific, chromosomes is required for the maintenance of the extinction of differentiated functions, the phenomenon of re-expression shows that *the sites of these stable epigenetic changes are either*

---

* Recalculation by Dr. Drew Schwartz of the data presented by Klebe et al.[136] has shown that the probability of involvement of chromosome C_8 is as high as that of chromosome C_{10} (personal communication).

The numbering of human chromosomes by Klebe et al.[136] is based here on the London Conference system.[154]

*in the chromosomes themselves or, if they are elsewhere—say in
the cell membrane—they are relayed to the chromosomes in order
to become effective* (Note 30).

I think that this conclusion is justified unless one assumes that
the quasi-general occurrence of extinction of differentiated functions has nothing to do with the normal processes of regulation
of these functions, but is simply the result of "messing them up"
by hybridization (Note 31). This appears to me quite unlikely
however, because most of the hybrids in which the extinction of
differentiated functions has been observed appear as perfectly
normal when judged by two other criteria, namely: (1) the joint
expression of both parental genomes, as witnessed by the co-
dominance of parental household markers; and (2) the co-ordinated replication of most, if not all, chromosomes of both parents.
There is no obvious reason for "messing up" only luxury functions, except if one assumes, as I did all along, that they represent
a separate class of cellular functions (distinct from household
functions), subject to a separate (epigenetic) system of controls.
But then, taking all this together with the fact that the cell
membrane is demonstrably rearranged by cell fusion (as witnessed
by the rapid intermixing of surface antigens in human-mouse
heterokaryons[75]) and with the accumulating evidence for the
involvement of the cell membrane in the expression of differentiated functions (p. 144), it appears to me more logical to
conclude that the extinction of luxury functions is not an
insignificant artifact but actually reflects the regulatory mechanisms which govern their expression, i.e. differentiation. For me,
the very occurrence of re-expression is the strongest argument in
favor of this view.

However, the final evaluation of the relevance of the extinction
of differentiated functions to the processes of normal determination will obviously have to await the production of viable hybrids
from crosses of normal diploid differentiated cells.

Possible mechanisms of extinction

When we first reported the extinction of melanogenesis in melanoma $\times$ fibroblast hybrids, we suggested that it may be due to
the synthesis by the fibroblast genome of a negatively acting diffusible regulator substance: the apparently total absence of

dopa-oxidase activity in the hybrid cells is indeed most simply explained by assuming that its synthesis is arrested by a repressor of the bacterial type[47, 48] (Note 32). However, there is no decisive proof as yet, *either in this or in any of the other cases* I talked about, that the extinction of a differentiated function is due to a block at the transcription level rather than at the level of translation. All facts thus far observed are *compatible with* an interpretation in terms of partial or total repression of transcription; *none is compelling.* Similarly, all of them *are compatible* with post-transcriptional control. Finally, it must be pointed out that in no case has it been shown as yet that extinction does not result from the synthesis of inactive, so to say "faulty" proteins. As I mentioned earlier (p. 80), the presence of a modified S-100 *could be* the explanation of the (apparently partial) extinction of this function in glial cell hybrids;* but I do not believe that it is likely to be a general explanation of the extinction of luxury functions. I may point out that such a hypothesis fits least well the extinction of hemoglobin synthesis in fibroblast-erythrocyte heterokaryons: on the assumption under consideration, Fe^{59} incorporation (into heme or hemoglobin) should have been detected indeed. In fact, the observations on these heterokaryons (to which, as it must have been noticed, I ascribe a particular importance because they concern the luxury function of a normal, diploid cell) clearly point to the operation of extinction at the level of transcription or translation (Note 33).

I shall add only that, should I be wrong and should it turn out that extinction is always due to the presence in the hybrids of "faulty" inactive proteins resulting from the modification of luxury molecules *after* their synthesis, this would imply that the corresponding messengers are correctly translated in the hybrid cells. This would indeed suggest that hybridization "makes a mess" of normal differentiation; however, by the same token, it would prove an extraordinary stability of the *differentiated state itself*; and we would again face the question whether this stability is a special feature of *neoplastic* differentiated cells which have been used in all of the experiments I have described. The answer to this important question will obviously have to await similar experiments on hybrids capable of multiplication, derived from

* See, however, Note 26.

crosses of normal, diploid differentiated cells. I expect such hybrids will soon be available.

Meanwhile, let's assume that the reappearance of luxury functions upon loss of chromosomes, which has now been observed in two different types of hybrids, proves to be correlated with the neoplasticity of the differentiated cells used in these experiments (Note 34). We shall then ask again whether the high stability of their differentiated state is due to changes in the chromosomes themselves or elsewhere in the cell,* in particular in the cell surface which, in neoplastic cells, is known to be different from that of their normal equivalents. I am sure you are all aware of the beautiful work done here, at Princeton, on the molecular basis of experimentally induced modifications of the surface properties of normal and malignant cells, and on the consequent changes of their social behavior (expressed by the presence or absence of contact inhibition of growth) which indubitably result from different transmission of external signals to their nuclei.[19, 20, 21] The genetics of these changes should, in principle, bring an answer to the question we are asking, and one approach to it is, of course, the study of the fate of malignancy itself and of the associated surface changes in hybrids between malignant and nonmalignant cells.

* The presence of viruses in most of these cell lines should also be kept in mind.

CHAPTER 7

Application of Cell Hybridization
to the Study of Cancer*

*It is surely true that no theory of cancer—or of biology—
is acceptable unless it comprehends neoplasia as one of
the possible consequences of biological organization.*
(Leslie Foulds. 1969, p. vi)

Since the discovery of somatic hybridization was made and its
generality established on pairs of cell lines of which at least one
was highly malignant (or transformed by an oncogenic virus), it is
not surprising that the malignancy of hybrids and the presence in
them of characteristics correlated with the malignant (or transformed) state were examined in the earliest investigations. All
publications on the subject† concurred in showing that the hybrid
cells, which were testable by inoculation into appropriate mice,
apparently inherit the high malignancy of the more malignant
parent (a few exceptions were observed and are mentioned in
Refs. 51 and 225). With the notable exception of hybrids with
fibroblasts of line 3T3[253, 254, 260] (which is exceptional among
permanent lines in that it is contact inhibited), all other hybrids
also appeared to inherit from the noncontact inhibited parent the
property of multilayered growth; and, in the case of virus transformed cells, the virus induced antigens.[50, 51, 254, 258]

All the hybrids referred to above (except those with 3T3[253, 254])
were the results of spontaneous hybridization, and were identified
karyologically. In a number of cases, additional evidence as to
their hybrid nature was provided by the co-dominant expression
of the parental histocompatibility antigens.[51, 61, 77, 217, 225, 239]
However, the karyological constitution of the tumors was not
examined thoroughly in these early investigations, and sometimes
not on the tumors themselves, but on cultures of explanted
tumors (see however Refs. 212, 225). Some of the conclusions

* See Note 35.
† See References 4, 6, 51, 77, 217, 225, 254, and 258.

arrived at in these studies therefore appeared as obsolete in the light of new results published in 1969 by Henry Harris, George Klein, and their co-workers.[104] Even though, as we shall see, they do not answer the question which led me to talk about cancer (p. 98), I must tell you about them because of the great potential importance of the observations of Harris *et al.* on the *suppression of malignancy* for the understanding of neoplasia and, hence, probably for that of normal regulatory mechanisms.

Suppression of malignancy

In a first series of experiments, hybrids between cells of three highly malignant mouse ascites tumors (Ehrlich, SEWA, MSWBS) and fibroblasts of the permanent line A-9,* possessing a low degree of malignancy, were tested for malignancy by inoculation into mice of the appropriate genotypes. The percent of takes showed that the malignancy of the hybrids is essentially similar to that of A-9 or, in other words, that the high malignancy of ascites cells was *suppressed* in the hybrids with A-9. Karyological analysis of three tumors which *did* develop following inoculation of the hybrids showed that they all had greatly reduced numbers of chromosomes (about 80 instead of the initial 128), and that the chromosome loss particularly affected the number of biarmed chromosomes contributed almost uniquely by A-9: the average reduction in the number of the latter was from 24 to about 4 to 10[104] (Note 36).

Table 7.1, taken from a later and more extensive publication[138] gives the chromosomal constitution of tumors produced by the inoculation of Ehrlich $\times$ A-9 hybrids. The data show that the tumor cells contain fewer chromosomes than the cells of the injected population; but unfortunately, the take incidence of the later is not indicated, and neither is the route of inoculation nor the kind of host used (adult? newborn? irradiated or not?) which seem to affect the percent of takes (see Note

---

* The three tumor cell lines used in these experiments were: (1) Ehrlich ascites, derived many years ago from a mammary carcinoma of a mouse of unknown origin; it shows very weak histocompatibility antigens and grows in *any* strain of mice; (2) SEWA, a tumor induced in a newborn A.SW mouse by polyoma virus; (3) MSWBS, a methylcholantrene induced tumor in an A.SW $\times$ A F$_1$ mouse. All three tumors are adapted to growth in ascitic form and the last two are characterized by the H-2^s histocompatibility antigen complex. The A-9 fibroblasts are an HGPRT− line of L cells,[151] derived from a C3H mouse and bear the H-2^k antigen complex.

TABLE 7.1. Chromosomal constitution of tumors produced by Ehrlich/A-9 hybrids[a]

Cell type	Total no. of chromosomes	
	Range	Mode
Ehrlich/A-9 hybrid cells injected	104–150	128
Tumor 1	83–91	84
Tumor 2	64–88	82
Tumor 3	72–96	89
Tumor 4	72–92	81
Tumor 5	80–98	88

[a] From Klein et al.[138]

36). Unfortunately also, the comparison of the data in Refs. 104 and 138 is rendered impossible by the lack of this information in Table 7.1.

A loss of chromosomes of similar magnitude has also been observed on very prolonged *in vitro* cultivation; but it did not affect the biarmed chromosomes in particular, and did not result in increased malignancy of the population as a whole. This indicates that the suppression of the high malignancy of the Ehrlich ascites cells is due to specific (although as yet unidentified) A-9 chromosomes, and that, in the inoculated animals, segregants are selected for which have lost these specific chromosomes and thus regain high malignancy (see however Refs. 65, 103, 180, and 212).

Later experiments showed that several other fibroblasts, derived from the L line, similarly reduce to their own level of malignancy, that of Ehrlich, SEWA, MSWBS, as well as of two other highly malignant tumor cells.[103, 138] *

In the very first report on these interesting investigations,[104] it was pointed out that the conclusion earlier arrived at by several authors[4, 5, 6, 50, 51, 225] that in somatic hybrids malignancy behaves as a dominant trait may have been erroneous because no thorough investigation of the karyological constitution of tumors produced

* The other two fibroblast lines are B-82 (TK deficient) of Littlefield and A9RI, a revertant of A-9 with restored HGPRT activity, isolated in Harris' laboratory. The two additional tumor cell lines are YAC and YACIR. Both are ascitic sublines of a lymphoma induced by Maloney virus in an A/Sn mouse and carry the H-2[a] antigen complex. YACIR has been selected for low concentration of Maloney surface antigen.

101

by inoculation of the hybrids had been made. (Parenthetically, it is rather amusing to remark that on this occasion Harris *et al.*[104] did not fail to give all [but one] relevant literature references!*) In my opinion, the most important experiments among those which led to this conclusion were those performed by myself in co-operation with Drs. Defendi, Koprowski, Scaletta, and Yoshida,[51] because they involved hybrids not between cells differing only in the *degree* of malignancy, but between highly malignant (polyoma transformed) and *normal* diploid T-6 fibroblasts. Although karyological analysis was performed directly on *some* of the tumors and confirmed their origin from the inoculated hybrid cells, no correlation between integrity of the tumor cell karyotype and malignancy was detected, probably because chromosome losses made the karyotypes of the hybrid populations, *prior to inoculation,* extremely variable (Note 37).

This is in fact the situation Harris, Klein, and their co-workers encountered when (realizing the importance of crosses of malignant cells not with cells of low malignancy but with normal diploid cells, on which I insisted in 1965,[61] and the possibility that the suppression of malignancy may be confined to L cells and their derivatives[104, 138]) they investigated the properties of hybrids between several types of malignant cells and normal fibroblasts.[15] Owing to this, the hybrids with L cell derivatives were unfortunately to remain the only ones providing clear evidence of suppression of malignancy.†

Hybrids between malignant and normal cells

The first hybrids of this sort obtained by Harris' and Klein's groups were between Ehrlich ascites tumor cells and diploid T-6 fibroblasts.[15] Their study proved disappointing: the hybrids regu-

---

* See References 65 and 103.

† Meanwhile, the first preliminary account of this work of Harris *et al.*[104] was accompanied by the following anonymous comment in *Nature*'s "News and Views": "These results with hybrids fly in the face of earlier reports that the fusion of a cancer and a non-cancer cell gives a malignant hybrid" (*Nature*, 1969, 223: 345–346). It went on to suggest, along with the authors of the article thus hailed, that "It may be that the ability of A-9 to suppress cancer is correlated with its ability to metabolize in the presence of 8-azaguanine. If so, this would open up a direct approach to the biochemical mechanisms involved in the suppression of malignancy." That this hope was vain was probable from the already existing literature, and soon proved by Harris and Klein themselves. (See Refs. 65 and 103.)

larly produced tumors in the inoculated mice. This is a result formally similar to those we obtained,[51, 217] but here it is ascribed to the (apparently nonselective) extensive chromosome losses (ca. 20%) which occur in the hybrids during their *in vitro* propagation *prior to* inoculation.

In the words of the authors themselves, "These observations on Ehrlich/fibroblast hybrids are, as far as the main point of the investigation is concerned, inconclusive. They do not decide whether a diploid fibroblast fused with a malignant cell can suppress malignancy in the hybrid; but they do show how instability of the chromosomal constitution of the hybrid cell may profoundly affect the interpretation of the results in such fusion experiments. In order to decide whether a normal diploid cell can suppress malignancy, it is obviously necessary to construct hybrids that maintain the complete chromosome sets of the two parental cells at least long enough to permit adequate assays to be made. In the following paper we present some observations on such hybrids."[15]

Accordingly, hybrids were isolated from two new crosses: the first one between SEWA and T-6 fibroblasts, the second between TA_3 (an ascitic form of a mammary adenocarcinoma) and diploid fibroblasts from AC A mice. Both tumor lines were chosen for this study because, contrary to Ehrlich ascites, they are characterized by low chromosome numbers [43 (42–44) in SEWA and 40 in (the "almost euploid") TA_3] and it was thought that the chromosome loss may therefore be less rapid in hybrids of these tumor cells with normal fibroblasts than in the Ehrlich $\times$ fibroblast hybrids.

Most of the SEWA $\times$ T6 hybrids (11 out of 16 clones) did not live up to this expectation: they showed considerable chromosome losses upon the earliest examination. The other 5 hybrid clones, however, still had at this time total chromosome numbers near to the expected ones. Inoculation of all clones with reduced chromosome numbers into irradiated syngeneic newborn mice resulted in 100% takes, but so did the injection of 2 of the 5 hybrid clones with the high chromosome numbers (Table 7.2). Examination of the data in this table shows that all of the tumors have chromosome numbers lower than those of the inoculated populations. However, the conclusion that in this case, like in

TABLE 7.2. Karyotypes of SEWA/fibroblast hybrids and of tumors produced by their inoculation into syngeneic hosts[a]

Cell type	Chromosome number in vitro		Number of takes[b]	Percent of takes	Modal chromosome number in tumors
SEWA	—		—	?[c]	43 (42–44)
T6T6	40		—	—	
Hybrids, expected	83	(82–84)			
Hybrids, observed:					
Clone 2	82	(76–86)	11/19	58	(63, 80)[d]
Clone 3	80	(76–90)	13/27	48	72, 73, 67, 85, 66 and 81–83
Clone 5	79	(76–85)	6/6	100	41, 69
Clone 13	82	(75–85)	16/31	51	62
Clone 15	81	(76–90)	3/3	100	e
7 other clones		60–75	66/68	~100	41–78

[a] Compiled from Tables 1, 2, and 4 in Ref. 262.

[b] Number of animals with tumors/total number of animals inoculated with 1.2×10^4 to 3.6×10^6 cells. Inoculations made subcutaneously into newborn irradiated A.SW $\times$ CBA F_1 hybrids.

[c] The tumorigenicity of SEWA under the conditions of inoculation of the hybrids described in [b] cannot be evaluated from the data in Refs. 104, 138, and 262. The only statement given is that, in ascites form, "an inoculum of 10^6 cells kills most animals in 3–4 weeks."[104, 138]

[d] The numbers in parentheses indicate modal chromosome numbers of tumors which arose in allogenic hosts.

[e] Not given.

that of the SEWA/A-9 hybrids, "the evidence indicates a very low level of malignancy for hybrid cells in which the complete parental chromosome sets are retained"[262] is, in my opinion, extremely weak owing to the karyotypic instability of the SEWA/fibroblast hybrids. In fact, as the authors state, "Since the tumorigenicity of the SEWA/fibroblast hybrids was profoundly modified by chromosome losses in vitro, populations of TA₃/fibroblast hybrids with *very high chromosome numbers** were selected for special study."[262] These hybrid cells with modal chromosome numbers 112, 114, and 117 respectively,† did indeed show low malignancy to begin with (cumulative take indexes in the range 5 to 20%), and the tumors produced upon their inocula-

* Italics are mine.

† These hybrids must have arisen from fusion products in which the chromosome sets of one of the parents had doubled, since the expected modal chromosome number in hybrids of this cross is 80 (cf. Note 10).

tion had chromosome numbers lower than those of the injected populations (mostly in the range 80–90).

Thus, although none of the crosses with normal fibroblasts gave results as clear as those of various tumors with cells derived from L cells, my impression is that, all in all, there most probably is a correlation between retention of chromosomes and suppression of malignancy even in the case of hybrids with normal fibroblasts; and that Harris, Klein, and co-workers have put their fingers on an important phenomenon.

Since, according to Klein, a most distinguished immunologist, this suppression of malignancy cannot possibly be due to histo-compatibility factors,* these observations suggest that, contrary to what I thought, *malignancy behaves as a recessive trait.*[262]†

Significance of suppression of malignancy

Turning to the significance of these facts, I shall start with a quotation from the paper by Wiener, Klein, and Harris:[262]

"We have used the term suppression of malignancy to describe the non-malignant character of the 'complete' hybrids resulting from the fusion of malignant and non-malignant cells. This term seems to us to be an accurate description of the phenomenon. Whatever heritable character is responsible for the malignancy of the tumour cells, it is not irretrievably lost as a result of cell fusion. All the hybrids we have tested are capable of generating malignant variants when certain chromosomes are eliminated. It is therefore difficult to escape the conclusion that in the hybrid cell the non-malignant partner contributes some factor that suppresses or holds in check the malignant properties of the parental tumour cell: and this suppression is removed when certain chromosomes derived from the non-malignant parent are lost. The frequency with which such malignant segregants are generated in the hybrid cell population appears to be related to the stability of the

---

* One may wonder, however, whether the possibility that hybrid cells carry new antigens (foreign to the inoculated hosts) is entirely excluded; and one is left wondering why the take incidences, in some cases, are not the same in syngenetic and allogenic mice even though all of them are newborn irradiated animals.[262]

† I have not discussed here the extensive work done by Klein's group on the antigens of many of the hybrids, the malignancy of which has been described above. For a first account of this work see Ref. 139.

chromosomal constitution of the hybrid cells. Those hybrids that undergo rapid and extensive chromosome losses *in vitro* give high take incidences when injected into the animal; those that tend to retain complete parental chromosomal sets for longer periods of cultivation *in vitro* give low take incidences.

"In Darwinian terms these results can be described very simply. The tumour cells used in the present study (and perhaps all malignant tumour cells) are the end products of selection: these cells have been selected for the ability to grow progressively *in vivo*. When such cells are fused with cells that have not been selected for this property (non-malignant cells), the malignant character of the tumour cells is not expressed in the hybrids. In order to obtain a subpopulation of malignant cells from these non-malignant hybrids, one must select again; and the frequency with which malignant subpopulations can be selected is a function of the degree of genetic variation in the hybrid cell population on which the selection operates. By far the most important source of this variation in the present circumstances appears to be the stability or instability of the chromosome set. In the light of the extensive observations that have been made on the role of cell selection in both experimental and human tumours all this, it seems to us, is hardly surprising.

"If our results are to be expressed in Mendelian terms, then one would say that malignancy behaves as if it were a recessive character. In our range of hybrids this is the rule; and we know of no decisive exception in any other studies. But there is no *a priori* reason why malignancy should invariably be recessive, and it may be that on further study of a wider range of material, cases may arise where malignancy behaves as a dominant character. For example, in a case where the non-malignant parent cell is susceptible to a particular oncogenic agent carried by the malignant parent cell, and this agent persists in the hybrid cell, one might expect that the hybrid would be malignant. But the demonstration that malignancy is dominant requires not only that the hybrid cells on injection produce tumours; it must also be shown that the cells in these tumours have the full chromosome sets of the two parent cells. In our material, this has not been found.

"We have now to consider what relevance the results of the present investigation have to the genesis of malignant tumours. One theory of the origin of malignancy would fit our findings very well; the idea that malignancy results from a specific genetic loss. If malignancy were the result of a genetic deletion, or of a defect that produced a non-functional gene product, then recessiveness of the malignancy in hybrid cells would be expected, for the non-malignant or normal partner in the hybrid would then be able to complement the defect. If more than one type of genetic loss could produce malignancy, one might expect that hybrids between different kinds of tumour cell might sometimes show complementation and thus be non-malignant. To explore this possibility we are at present examining the behaviour of hybrids produced by fusing together different kinds of malignant tumour cell."

I shall limit myself to pointing out that, while the observations of Harris, Klein, and co-workers on the suppression of malignancy are indeed compatible with the idea that malignancy is caused by a recessive chromosomal lesion (mutation or deletion of genetic material in the malignant cell), they are strikingly reminiscent of the extinction of luxury functions described in Chapter 6. The phenomenological similarity between suppression and extinction makes one wonder whether malignancy is not, after all, due to epigenetic changes rather than to genetic ones, *sensu stricto*. This similarity is especially striking if one recalls (1) the now available indications that both extinction of luxury functions (cf. Note 29) and suppression of malignancy are markedly affected by the dosage of the relevant genes (the latter is suggested by the recent studies of Okada and co-workers on the tumor forming capacity of a series of hybrids obtained by reciprocal double hybridization of Ehrlich ascites and L cells,[191] * as well as by the work briefly described above on the suppression of malignancy in the TA_3/ fibroblast hybrids which were selected for study *on the basis of*

* The tumor forming capacity is found to be in the order: $E > LEE > LE > LLE$ (where E stands for an Ehrlich ascites cell and L for an L cell).

high chromosome numbers)[262] (Note 38); (2) the re-expression of luxury function upon loss of chromosomes now established in several types of hybrid cells (pp. 90f. and 93f.) which is formally analogous to the reappearance of malignant segregants upon loss of chromosomes by nonmalignant hybrids between malignant and "nonmalignant" cells described by Harris, Klein, and their co-workers (see above, pp. 100ff.); (3) the recent observations of Silagi and Bruce[226] on the (apparently reversible) considerable reduction and eventual abolition of malignancy of a mouse melanoma by exposure to BUdR, a substance known to reversibly interfere with the expression of differentiated functions in many normal and malignant cell types (for the literature on this subject, see Ref. 226).* From these observations, Silagi and Bruce concluded that, "The suppression of cytodifferentiation in these melanoma cells and embryonic cells, coupled with the modification of malignancy, leads to the hypothesis that both differentiation and malignancy are regulated by similar cellular mechanisms." (Similar views have been previously expressed by Braun,[14] Pierce,[198] and Markert.[160])

I was recently told by Harris and Klein that the inference that malignancy behaves as if it were "the result of a genetic deletion, or of a defect that produced a nonfunctional gene product"[262] finds support in the as yet unpublished fascinating observation that crosses between two different highly malignant tumors (YACIR × MSWBS) produce some nonmalignant hybrids (A. Cochran, F. Wiener, G. Klein, H. Harris, personal communication). But these results appear to me equally compatible with an hypothesis ascribing malignancy to epigenetic changes (Note 39).

Returning to the question from which we started (i.e. the possible *genetic correlation* between malignancy and cell surface changes, p. 98), I note that in their publications,[15, 138, 262] Harris, Klein, and co-workers say nothing about the contact behavior of

---

* It is notable that the effect of BUdR is constantly associated with pronounced changes in the morphology of the cells and with their adhesiveness to glass and/or different plastic surfaces, which undoubtedly reflects changes in the properties of the surface properties of the cells. While the mechanism of action of BUdR remains a matter of debate, Schubert and Jacob[221] present experimental evidence that (1) BUdR does not act by virtue of incorporation into DNA and (2) that it probably affects the synthesis of carbohydrate moieties of the glycoprotein and mucopolysaccharide containing components of the cell surface.

the different hybrids; I suppose this is so because the malignant parents of the hybrids they studied grow only in suspension. I am sure that new hybrids of the sort needed will soon be produced, and information obtained on this important question.

CHAPTER 8

Concluding Remarks and Postscript

> *When we turn to look at the nature of biology itself, we*
> *see stretching before us an almost unlimited number of*
> *important, interesting, and unsolved problems. This is*
> *partly due to the inherent complexity of biology and partly*
> *due to a passionate desire to understand the world around*
> *us and our own natures in particular. There are so many*
> *things we should like to know—and like to know about in*
> *considerable detail—that we need not seriously worry at*
> *this stage that the subject will become exhausted.*
>
> (Francis Crick. 1970, p. 614)

In the "Introduction" to these lectures, I expressed the hope that, after you will have heard what I have to say about the present status of some biological problems, you will share my belief that there is still plenty to do for biologists of the present and next generations. It must be clear to you now that among all the problems to which I alluded, those that I had foremost in mind were problems of development, to which I devoted the greater part of my lecture time. It must have become clear to you also that, at the moment, we have no answers even to some of the most elementary questions raised by the last and most accessible stage of development, i.e. differentiation *sensu stricto*, let alone determination and morphogenesis.

I do believe that studies on hybridization of differentiated cells have brought to light a few facts (extinction of luxury functions, relative independence of the controls of differentiated functions of one "set," re-expression of "extinguished" luxury functions, apparently upon loss of definite chromosomes) which may represent new and effective leads for the analysis of the mechanisms of differentiation and determination. Among these new facts, that of re-expression of luxury functions, which bears witness to the ability of different epigenetic states to be propagated over numerous cell generations without being expressed, appears to me to be of the greatest importance and to hold the greatest promise even though its intimate mechanism is for the moment entirely enigmatic. Its very occurrence has, in my opinion, already settled the major alternative [which I stated when I spoke of my articles of faith (Chapter 4)]: inheritance of *intrinsic* differences between individual, differently determined cells (postulated by me all along) *vs.* group stability based on reinstruction (Grobstein's notion). Re-expression reveals, as we have seen, the persistence (i.e. inheritance) of the epigenotype of the "differentiated parent" in hybrids in which the luxury functions of the latter had been "extinguished." The fact that re-expression is a *rare event occurring in single cells embedded within dense populations of different phenotype* (see, for example, the conditions under which re-expression was obtained in hepatoma $\times$ BRL hybrids, p. 89), precludes the applicability of the notions of group stability and reinstruction (which are made also most unlikely by the correlation between re-expression and change in chromosomal constitution). I put particular hope in the further exploration of the phenomenon of re-expression because I believe it may contribute to the solution of the often raised problem of the respective roles of the chromosomes and of the cell membrane (and of their interactions) in determination and differentiation. You will recall that re-expression of luxury functions is brought about by loss of chromosomes of the "extinguishing" parent and as pointed out (pp. 89 and 95), is constantly correlated with changes of cell morphology and/or cell behavior, undoubtedly due to changes of the cell surface.* Notice, however, that (1) the reverse is not true: while all segregated hybrids show a return to the morphology of the parent whose chromosomes are retained, only a fraction of them (presumably that which has lost certain specific chromosomes) re-expresses a given function; (2) moreover, two different luxury functions of the same parent are apparently re-expressed independently of each other (p. 91), presumably because they depend on the loss of different chromosomes. If the cell membrane really plays a specific role in the regulation of the different luxury functions, the gross changes or cell morphology, and/or behavior which leap to the eye must probably involve much more subtle,

---

* See also footnote on p. 108 and Notes 29 and 38, where surface changes, associated with changes in gene expression observed in other situations are described.

different and specific changes in the composition of the cell membranes of segregated hybrids differing with respect to re-expression or nonre-expression of different luxury functions. The existence of such subtle specific changes should soon become demonstrable by the various techniques which are beginning to be used for "mapping" cell membranes (see for example, the work of Boyse, Old, and their group[2]); and should permit the establishment of correlations between re-expressed luxury functions and presence or absence of definite components of the cell membranes on the one hand, and between the latter and the presence (*vs.* absence) of definite chromosomes on the other. Such studies may take us far beyond the area of elementary embryological problems to which I limited most of my theoretical remarks (in Chapter 4). What I have in mind now are the much wider problems of the genetic basis of morphogenesis in multicellular organisms; of the role therein of the relationships between different types of cells, mediated by the differential properties of their (tissue-specific) surfaces as the instruments of recognition of cell type; and of the genetic origin of these differences, i.e. their dependence on the nucleus *vs.* their genetic autonomy (Note 40).

It is because of the importance of these problems (which are high on Crick's[37] and Hershey's[110] lists of fundamental and as yet unresolved problems of molecular biology—if indeed "molecular biology can be defined [as Crick[37] suggests] as anything that interests molecular biologists") that I thought it was important to give you a clear picture of the nature of the observations which resulted in the new (and, in my eyes, particularly significant) type of evidence for the stability of the epigenotype of determined cells.

On the other hand, I deliberately refrained from suggesting a concrete mechanism for the explanation in molecular terms of epigenetic changes. For one thing, using Socrates' words, I considered that my role here was that of a midwife: it was not to give you *my* ideas, but to help you deliver *your own*. For another (and more serious one), while I have little doubt that an interpretation of stable epigenetic states based on models of "circuitry" of micro-bial type could be constructed, I feel that to try to force it into such a scheme would be not only premature but definitely dangerous at this stage: in the field we are considering, the time has come to cease satisfying ourselves (as is still too often done) with what may be mere analogies which, almost inevitably, result in too rigid adherence to what has become known as *The Dogma*. While there is no doubt in my mind that the latter represents the greatest achievement in biology since the Darwinian revolution, I feel we must not loose sight of the possibility that, based as it and especially *its corrollaries* are on the study of bacteria and viruses, they may have to be supplemented by additional principles which may have evolved concomitantly with the emergence of more complex organisms with very different requirements with respect to regulatory mechanisms (of which gene amplification, if demonstrated for other than ribosomal genes, may be an example capable of accounting both for the establishment and the stability of epigenetic states).

Lastly, and most importantly, I do not want you to overlook that even though it has been said that since the discovery of genetic regulation in microorganisms, "biochemical differentiation (reversible or not) of cells carrying an identical genome, does not constitute a 'paradox,' as it appeared to do for many years to both embryologists and geneticists,"[174] this statement tells us nothing about the nature of the primary causes responsible for the orderly, divergent biochemical differentiation of different cell lineages derived from a single egg (whether its mechanism be based on self-maintaining regulatory states or, for that matter, on any mechanism of differential gene activation or amplification). *The real problem is that of the origin (seat) of the asymmetrical causes which bring about these asymmetrical effects.* And that is why I would like to quote, before I conclude, a few paragraphs from the penetrating paper by Hershey on "Genes and Hereditary Characteristics"* in which he puts his finger on precisely this point:

> "The discovery of early geneticists that a unique set of genes
> determines the characteristics of the individual at once raised
> the question whether or not heritable characteristics are deter-

---

* Hershey's paper in *Nature*[110] is, in fact, a version of the first part (devoted to problems of development) of his Annual Report, as director of the Genetics Research Unit, Carnegie Institution, which appeared under the same title in Ref. 109. The two versions differ slightly in wording and emphasis, and some quotations have been taken from each of the two papers.

mined exclusively by those genes. This question remains un-answered and is, in fact, difficult to phrase intelligibly. The demonstration by molecular biologists that a linear genetic code can be translated into three-dimensional structure, as in the assembly of virus particles, showed that in principle the known mechanisms of inheritance could be the only mechanisms. (The example of the viruses is important because there one can observe the regeneration of quite different viral species in the same cellular milieu, depending only on the kind of DNA molecule introduced at the start.) The inference that all three-dimensional structure is encoded in nucleotide sequences does not necessarily follow, however. I shall call that inference the unwritten dogma, for it must be shared at least by those biologists who consider molecular biology all but finished."[109]

Turning "to some problems of cellular organization that seem to lie beyond the reach of current molecular principles," Hershey remarks that "the successes, notably enzymology and genetics, should not be allowed to overshadow the failures, such as analysis of cellular development."[110] Here, elaborating on the traditional view (as expressed by Tartar[244]) that, "Guidance of the elaboration of formed parts in the cytoplasm is to be sought in neither a nucleus of unprescribed location nor a flowing endoplasm but in the most solid portion of the cell, namely, the ectoplasm," Hershey notes: "This simplifying hypothesis divides the cell into two parts: on the one side, genes and gene products; on the other side, a cell cortex or skeleton that seems to manifest supramolecular properties."[110] He then comments, "Strictly speaking, the hypothesis says nothing about the origin of the guidance system, which could in principle be recreated every time an egg is fertilized, and whenever a protozoon emerges from its cyst. But some sort of guiding principle is usually assumed. For example, Luria* remarked, in effect, that just as biological evolution is separate from cultural evolution, so is DNA evolution separate from the evolution of cellular organization." And Hershey adds: "This I suspect is as close as it is possible to come to a contemporary statement of the cell theory. Like all good theories, it admits alternatives: for example, the unwritten dogma, according

* My Ref. 156.

to which biological evolution is solely the evolution of nucleotide sequences"[110] . . . "The hypothesis of bipartite inheritance arose naturally, I think, from the fact that cellular differentiation can be reversible or not to any degree. The reversible changes seemed to imply extragenic determination and the irreversible changes suggested extragenic inheritance, that is, differentiation dating from the origin of the species. The force of the argument is now weakened by the evidence that developmental changes sometimes can be explained in terms of selective control of gene action.* But not all persistent changes in cell structure and behavior can be so explained."† What Hershey refers to here are chiefly experiments on morphogenesis and regeneration of cortical patterns in *Stentor* and *Paramecium* which brought to light "a primordium in the cell cortex that is indispensable to development and, at least in *Paramecium*, persists through both sexual and asexual reproduction,"[109] as well as the fact that polarity (both in *Stentor* and in *Paramecium*) resides in all parts of the cortex. "Perhaps, as many people think," remarks Hershey, "polarity represents something that was invented only once and has evolved since on its own."[110]

Turning to the well-known experiments of Hämmerling[96, 97] on interspecific grafting in the monocellular alga *Acetabularia*, Hershey notes that "Cortical inheritance, if it exists in *Acetabularia*, is apparently not species specific."[110] This leads to the inference that "Perhaps, contrary to what many people think, polarity resides exclusively in the gene-determined structures of polar molecules, and only genes evolve. If so, the only cell theory worthy of the name is wrong."[110]

And he concludes: "In *Stentor* and certain other cells, all reasonably large pieces of the cell cortex are equipotent with respect to regeneration of cortical patterns—patterns that are, moreover, subject to metastable variation. In bacteriophages, supramolecular patterns do not persist as such but recur, apparently residing exclusively in gene determined structures of individual molecules. *Acetabularia* presents a tantalizing example somewhere in between. Taken together, the facts encourage us to see in cortical polarity a historical invention that ought to be analyzable in

* My Ref. 233.
† My Refs. 233 and 7.

115

terms of structure and process. They do not encourage us to think that the task of molecular biology is finished, even at the cellular level. In Tartar's words,* *'Our greatest lack and most fruitful opportunity in biology lies in conceiving and testing the nature and capabilities of persistent supramolecular patterns'* "† [110] (Note 41).

POSTSCRIPT: In rereading the text of my lectures, these words come to my mind:

I have the feeling that I have expressed myself . . . not merely somewhat longwindedly but also rather sharply. This observation may serve as my excuse: one can really quarrel only with his brothers or close friends: others are too alien [for that].

(Albert Einstein. 1949, p. 688.)

* My Ref. 244.
† Italics are mine.

Appendix

Note 1. Among other important areas, not covered in my lectures, which have profited from the use of cell hybridization, virology must be mentioned in the first place. Particularly important examples are its use in the study of (1) virus rescue, first clearly demonstrated by Koprowski *et al.*[142] and Watkins and Dulbecco[250] for SV_{40} and now extended to other viruses by numerous workers; (2) the integration of oncogenic viruses in transformed cells (reviews in Refs. 253 and 254); and (3) factors affecting cellular susceptibility to virus infection (review in Ref. 83).

As will be seen from brief references in Chapter 7 and Note 31, somatic hybrids are being actively used for immunological research.

In the domain of macromolecular syntheses, most interesting results concerning regulation of the synthesis of ribosomes have been obtained by Green's group.[56, 241]

Lastly, Sendaï induced cell fusion is beginning to be used in embryological studies.[78]

Note 2. I may mention that in the first *in extenso* paper by Harris *et al.*[107] on the formation of heterokaryons by the action of UV irradiated Sendaï virus, it is stated that the idea to produce artificial *heterokaryons* can be traced back to the middle of the last century and a number of references are given to observations made, chiefly by pathologists, on the occurrence of polykaryocytosis (i.e. formation of multinucleate cells) in inflammatory reactions. However, all these references are totally irrelevant because there is every reason to suppose that all the examples quoted relate to the fusion of *genetically identical* cells and therefore, by definition, result in the formation of *homokaryons*. The only relevant mentions of earlier isolation of hybrid cells[5, 6, 66] are buried amidst references to old observations on the occurrence of bi- and multinucleate cells (i.e. probably homokaryons) in tissue cultures and on their probable viral aetiology.

117

Note 3. Explants of many normal (embryonic or adult) tissues *in vitro* give rise to propagating populations of normal *diploid* cells. Such populations, designated as cell *strains,* although they can usually be cloned, grow only for a finite number of generations, the number of which depends on the species of origin (*ca.* 20–30 for mouse cells, and *ca.* 50 for human cells). At the end of this time they are said to become "senescent": their growth rate declines and they cease dividing. In some cases, foci of rapidly dividing cells appear, which are capable of indefinite multiplication and give rise to what are designated as *permanent lines.* The karyotypes of these cells are usually abnormal (heteroploid) and comprise most frequently a subtetraploid number of chromosomes, some of which have undergone structural rearrangements. The karyotype of a given permanent line is generally variable and is described by the modal number and the range of variation. Since the modal number does not correspond to a simple multiple of a normal diploid chromosome set of the species (2*N*), it is referred to as the stemline number, abbreviated 1*s.*

Note 4. Karyotypic analysis of somatic cells is performed on preparations of (usually "blocked") metaphases. At this stage, each chromosome comprises two chromatids joined at the centromere, and it is the position of the latter along the length of the chromosome which defines the two major classes of chromosomes: the metacentric (bi-armed) chromosomes and acro- or telocentric chromosomes (those with a terminal centromere). Since at this stage the two chromatids are joined by the as yet undivided centromere, the telocentric chromosomes appear as V-shaped, the metacentric ones as X-shaped.

Further subdivisions are according to the more or less central location of the centromere, and the relative lengths of chromosomes and chromosome arms (example: submetacentrics, subtelocentrics, etc.).

Note 5. These heterokaryons were qualified as viable because they continued to metabolize for up to 15 days and in spite of the fact that most of them never underwent mitosis. This use of the term "viable" appears to me confusing and I prefer to limit its use, like in microbial genetics, to cells able to undergo at least enough divisions to form a visible colony (*ca.* 10^3 mammalian cells).

Note 6. A notable exception is the very profitable use of heterokaryons in virology to which references were given in Note 1.

Note 7. It is rather obvious that Sendaï virus causes cell fusion by an as yet unidentified effect on the cell surface[101] [in fact, it has been suggested (cf. Ref. 101) that "spontaneous" hybridization may be due to viruses carried by the parental cells]. Starting from a hypothesis formulated by Lucy[155] concerning the structure of cell membranes and the mechanism of action of factors causing membrane fusion, Pool *et al.*[204] were led to test the fusion inducing capacity of lysolecithin and indeed obtained fusion of cells of the same and different types. The possibility of substituting lysolecithin for Sendaï virus for the production of somatic hybrids would certainly present some technical and possibly theoretical advantages. In a recent paper, Croce *et al.*[38] report encouraging results.

Note 8. The reason(s) for this low yield of viable hybrids remain unclear. In fact, the mechanism of nuclear fusion itself is still open to question. In their very first publication, Harris and Watkins[106] reported that in HeLa $\times$ Ehrlich heterokaryons produced by the use of UV inactivated Sendaï virus, cell fusion is often followed by nuclear fusion. The latter was ascribed to fusion of interphase nuclei. In a later publication,[107] emphasis was shifted from this mechanism to coalescence of parental chromosome sets during mitosis, and in Harris' recent books[100, 101] fusion of interphase nuclei is no longer mentioned.

It appears to me highly probable that mononuclear hybrids containing genomes of both parents can be formed during synchronous mitosis of the parental nuclei with the formation of a single (regular or multipolar) spindle. That fusion of interphase nuclei also occurs cannot be excluded.

One of the first possibilities that comes to mind to account for the low fraction of diheterokaryons which give rise to viable hybrids is that the nuclei of binucleate cells formed by the fusion of cells in different phases of the cell cycle fail to achieve synchrony and hence either do not reach mitosis or undergo abnormal mitosis. However, a study by Rao and Johnson[207] of the results of fusion of synchronized populations of labelled and unlabelled HeLa cells, blocked at different stages of the cell cycle, does not appear to support this hypothesis. In fact, it revealed a surprisingly high degree of synchronization, and of normal mitoses, whatever the stage of interphase nuclei of the fused cells. Asynchronous mitoses were found regularly only when one of the partners was a mitotic cell and the other an interphase cell. Particularly important anomalies ("premature chromosome condensation" of the authors,[125, 127] termed "chromosome pulverisation" by Kato and Sandberg[133, 134]) result from fusions between mitotic cells with cells in $G1$ or S. (For some further details of Rao and Johnson's work, see Note 10.) However, the presumably low frequency of such combinations in mixtures of nonsynchronized cells does not favor the view that this is the main cause of the low yield of viable hybrids from diheterokaryons. It must not be forgotten, however, that (1) the conclusions from the quoted studies are based on microscopic observations of only the first post-fusion mitoses and (2) the viability of the progeny of the fusion products has not been tested. (Concerning the first point, it must be kept in mind that, as Johnson and Rao state,[126] "Perfect synchrony of nuclear events in a multinucleate cell demands simultaneous passage from a pre-DNA synthetic phase $(G1)$ into the synthetic (S) phase, followed by identical rates of synthesis in each nucleus and perfect coordination of passage into the premitotic $(G2)$ phase. Finally, the nuclei should enter prophase together and pass through the stages of mitosis (M) in synchrony. Such a perfect synchrony is unlikely in any multinuclear system." It is uncertain that cytological observations, as practiced by the authors, are sufficient to guarantee the "perfect synchrony" which may be required for the formation of viable hybrids.)

Furthermore, it must be added, on the one hand, that the cause of the nonviability of the majority of diheterokaryons may reside in events more remote (in time) from those observed at the first metaphase, and possibly not even directly related to events revealed by the usual karyological analysis (see Note 10); and, on the other hand, that it may have nothing to do with the mechanisms envisaged above and be due to quite different factors, such as the "induction" of latent viruses for example, some kind of interaction between newly formed hybrids the occurrence of which one is led to suspect from the peculiar relationships between the ratio of parental cells and the effective mating rate mentioned on p. 14, etc. Lastly, it remains possible that it is due to our "technical inadequacies" which probably account for the low plating efficiency of many cell types.

Note 9. Although this paper describes the occurrence of much more pronounced chromosome losses in some hybrids between cells of Littlefield's mouse cell lines A-9 and B82 after very prolonged culture (over 10 months) in HAT, no proof is given that these cells are viable. In a publication presently in press[57] (kindly communicated to me by Dr. E. Engel), it is shown that clones with low chromosome numbers can be isolated from such populations. In fact, the authors apparently succeeded in obtaining, after 27 months of culture in HAT, some clones with karyotypes quantitatively and qualitatively very similar to those of the parental cells (i.e. clones which had lost nearly 50% of the chromosomes of the original hybrids). It thus appears that, on *very prolonged* cultivation, (some) intraspecific hybrids can undergo chromosome losses comparable *in magnitude* (although not in rate) to those which are observed in some interspecific crosses.

Preliminary observations indicate that the chromosome losses observed by Engel *et al.* may be much more pronounced in HAT than in the same medium minus aminopterin (HT medium). This effect (action of aminopterin) is being investigated on numerous A-9 $\times$ B82 clones maintained in parallel in HT and HAT (personal communication of Dr. E. Engel).

Note 10. In connection with the question of the mechanism of preferential loss of chromosomes of one species in some interspecific hybrids, the following may be recalled:

In a speculative paper written soon after the discovery of the first viable mononucleate hybrids resulting from the fusion of cells of different species,[70] often characterized by very different generation times, Ephrussi and Weiss[71] remarked that the very viability of these hybrids points to the existence of nonspecies specific signals coordinating the initially different cell cycles of the parental cells. It is satisfying to observe that the existence of such signals has now indeed been confirmed by the studies of fusions between synchronized populations of HeLa cells referred to in Note 8. Briefly, the studies of Rao and Johnson[125, 127, 207] on heterophasic HeLa homokaryons have revealed the following effects: (1) shortening of $G1$ and S in the presence of $G2$ nuclei; (2) retardation of entry into mitosis of $G2$ nuclei in $G2/G1$ and $G2/S$ combinations (the delay lasting essentially until the S or $G1$ components are "ready" for mitosis); (3) "premature chromosome condensation" (PCC,[125, 127] also referred to as "chromosome pulverisation"[133, 134]) in $G1$, $G2$, and S nuclei in homokaryons resulting from fusion with mitotic cells. All of these effects are dose dependent (i.e. dependent on the relative proportion of nuclei of the two types in multinucleate homokaryons), operate *via* the cytoplasm, and appear to show a rather low degree of species specificity (clearly demonstrated for PCC). They are very similar to the effects observed by numerous authors in experiments on nuclear transplantation in a number of organisms. (The relevant extensive literature on the subject is covered in a recent review by Johnson and Rao[126].) The molecular nature of the synchronizing signals remains unknown, but some evidence points to their protein nature (and cytoplasmic origin),[126, 166] a notion contrasting to the view of Harris (cf. Ref. 100) on the rather trivial nature of the factors (changes of osmotic pressure and of movement of electrolytes) which are thought to bring about nuclear reactivation through changes of the state of condensation of nuclei in heterokaryons, and of their consequent activity (see Ref. 166 and Note 22).

Ephrussi and Weiss[71] also pointed out that nuclear synchrony is not achieved in some interspecific hybrids and suggested that, at first sight, this may be the cause of the preferential loss of chromosomes of one species observed in some of these hybrids, of which the most extreme examples known at the time were the human $\times$ mouse hybrids which rapidly lose most of the human chromosomes.[259] It may be recalled that in many of the human $\times$ mouse hybrids, preferential loss of human chromosomes is so rapid that for a long time it could be questioned whether these hybrids resulted from total fusion of the parental nuclei. It was, however, shown later that even the very incomplete human × mouse hybrids, must arise from total fusion of parental cells and nuclei.[122, 123] (The same appears to have been demonstrated for Chinese hamster × human hybrids[132] which also lose the human chromosomes with extreme rapidity.) However, it has been noticed also[122, 123] that even though the presence of the essentially complete genomes of both parents can be identified in the majority of *first* metaphases of human × mouse hybrids, approximately 25% of the latter already lack some human or both human and mouse chromosomes, and that the proportion of "segregated" hybrid cells increases in the course of the following days (that what is designated as first mitoses are indeed the first post-fusion divisions is rendered probable by the observation, in karyological preparations made very soon after virus induced fusion, of metaphases where the chromosomes of the two species can be recognized as forming two essentially separate groups on an apparently single spindle[122, 123]). Although it cannot be entirely excluded that the presence, at this early stage, of metaphases lacking many of the human chromosomes is due to artifacts, it appears much more probable that it is due to what we[71] described as "failures of coordination" which may extend beyond the first post-fusion mitosis. It is clear that purely fortuitous factors, such as irregular chromosome distribution owing, say, to the formation of multipolar spindles at the first division(s) (i.e. due, in turn, to failure of regulation of centrosome behavior) may have far reaching and protracted consequences (*via* the generation of daughter cells with different balances of many specific genes) on the further evolution of the clones to which they give rise and within which intense selection of the more balanced hybrids complicates the search for *the cause* generating this unbalance.

As to the nature of the "failures of co-ordination" resulting in the preferential loss of chromosomes of one species, one of the first hypotheses which comes to mind is that of a breakdown, in these interspecific hybrids, of the synchronization of the cell cycles which appears to be so perfectly achieved in the

HeLa *homokaryons* described in Note 8. This breakdown could result, for example, in the incomplete replication or condensation of the human chromosomes at the time of the first mitosis of the dikaryon (or "newborn" hybrid). However, as pointed out by Ephrussi and Weiss,[71] this hypothesis alone cannot account for the fact that the preferential loss concerns always the chromosomes of the same species. It appears to be difficult to account for this constancy without invoking "cell specific" interactions, an extreme case of which has already been mentioned (p. 58): I am referring to the HeLa + Ehrlich heterokaryons in which, according to Johnson and Harris,[124] the Ehrlich component almost totally inhibits DNA synthesis by the HeLa nuclei. Such a view finds support in the fact that some aspects of nuclear behavior are very different in crosses between cells of different lines of the *same two species*, for example human and mouse. Thus: (1) The rate of loss of human chromosomes is highly variable depending on the particular mouse and human cell lines chosen (see for example Ref. 254); (2) While in some human $\times$ mouse crosses the majority of hybrids comprise a single mouse genome ($1s$) + a few human chromosomes and a minority carry $2s$ mouse chromosomes, in other crosses the relationship is strictly the inverse, the majority of hybrids containing 2 mouse genomes (for references, see Jami *et al.*[122, 123]). (Notice that the origin of the "2-s-ness" of the mouse genome in these hybrids—from multiple fusions or from an extra-round of replication of the mouse chromosomes prior to the first post-fusion mitosis—is still a matter of debate;[122, 123, 208] and that while such an extra-round of chromosome replication in $G2$ nuclei has not been observed in Rao and Johnson's studies on heterophasic HeLa homokaryons, its occurrence in (some?) interspecific hybrids cannot be *a priori* excluded); (3) A possibly significant fact which will have to be taken into account is the recent finding that in the cross between mouse cells of line Cl.1D and cells of the human (SV_{40} transformed) line VA-2, some hybrids are formed in which the loss of chromosomes is "reversed": a few hybrids retain a more or less complete ($1s$ or $2s$) human chromosome complement while they lose the majority of mouse chromosomes.[122, 123]

All of these facts are difficult to fit into a scheme which

ascribes a major (if not unique) role in the preferential loss of chromosomes to the lack of synchronization of the nuclei in diheterokaryons.

Following their study of human × hamster hybrids, Kao and Puck[132] suggested that it is the slower growing parent of a hybrid that undergoes chromosome loss. This generalization does not seem to find support in (unpublished) observations of several investigators (personal communications).

Some alternative hypotheses have been formulated in our already quoted speculative paper[71] [for example, the possible role of hybrid (membrane and/or enzyme) molecules; non-recognition of regulatory signals; difference in timing of responses of the two parental genomes to some critical and appropriately recognized signal]. A number of other hypotheses (such as the unequal affinities of chromosomes of the two species crossed for some locus on the nuclear membrane) could be added at present. The above remarks suffice however to show that what is lacking for the choice between the different hypotheses and for the formulation of new ones is more information—of the kind obtained by Rao and Johnson's studies, but extended both beyond the stages thus far analyzed by them and to interspecific combinations.

Note 11. A new system for the selection of human × mouse hybrids has been reported recently by Kusano *et al.*[146] It relies on the use of mouse fibroblasts (3T6-DF8) lacking the enzyme adenine phosphorybosyl transferase (APRT). These APRT$^-$ cells are killed in medium (AA) containing adenine and alanosine, an antibiotic which blocks the *de novo* synthesis of adenylic acid, while "wild type" cells from which 3T6-DF8 was isolated, do grow in this medium. Both "spontaneous" and virus induced hybrids between 3T6-DF8 and normal diploid fibroblasts have been easily isolated by the use of AA medium owing to the retention of the human chromosome carrying the human APRT$^+$ gene. (The loss of the majority of human chromosomes has, in fact, enabled the authors to make the first steps in the assignment of the APRT gene to a human chromosome by a method essentially similar to that used for the localization of the human TK gene described in Chapter 3.

Successful isolation of hybrids between two Chinese hamster cell lines with increased resistances to Actinomycin D and Amethopterin, respectively, in a medium containing both drugs, has been reported by Sobel *et al.*[232]

Note 12. "Inborn errors of metabolism" have been used either to facilitate the selection of interspecific hybrids involving human diploid parents or to identify the activity of the human chromosomes in such hybrids. The first use of mutant diploid cells was made by Migeon and Miller[167] who isolated, in HAT, hybrids between cells of the TK⁻ mouse line Cl.1D and diploid cells from a human male with the Lesch-Nyhan syndrome, known to be HGPRT deficient.

Silagi *et al.*[227] obtained hybrids between cells of the HGPRT⁻ heteroploid human line D98/AH-2 (derived from HeLa) and diploid human cells from an infant with orotic aciduria, a rare autosomal recessive disease correlated with deficient activity of the two final enzymes in the biosynthetic pathway of uridilic acid. The hybrids were isolated by the semi-selective method using HAT.

The first hybrids between two human diploid cells carrying different X-linked enzymatic defects were produced by Siniscalco and co-workers,[229, 230] also by a semi-selective technique; but, in so far as I know, pure hybrid clones have thus far not been isolated. (For a recent review of the subject, and summary of interesting observations on the expression of X-linked genes, see Siniscalco.[231])

Lastly, I may mention the recent paper by Nadler *et al.*[183] in which the production of hybrids is reported by fusion of pairs of human diploid cells from different patients with galactosemia. Pure hybrid clones do not seem to have been isolated by the authors (see Note 24).

Note 13. Hybrids between green monkey and mouse (SV$_{40}$ transformed) cells are reported to lose rapidly and preferentially most of the monkey chromosomes.[27]

Hybrids between cells of Chinese hamster and of kangaroo-rat lose preferentially (and probably rapidly) most of the chromosomes of the latter species.[121]

126

Note 14. In a recent report, Pontecorvo[203] describes the results of ingenious experiments aimed at the *directed* chromosome elimination in somatic cell hybrids which should prove useful in many areas of research using these hybrids. The first experiments reported were performed on hybrids between TK⁻ 3T3 mouse fibroblasts × HGPRT⁻ Chinese hamster cells (line wg/3) which, normally, have a very stable karyotype. Irradiation with an appropriate dose of X-rays of *either one* of the parental populations *prior to* virus induced fusion, results in the preferential loss from the hybrids of the chromosomes of the irradiated parent. With a dose of 600 r, there is loss of up to 9 of the expected 66 3T3 chromosomes, and up to 8 of the expected 22 Chinese hamster chromosomes. (It appears however from Pontecorvo's report that the details of this technique will probably have to be adjusted for every new cross to which it will be applied, for it is shown that an excessive dose of irradiation, while it induces more extensive chromosome losses, not only produces more chromosome aberrations but, moreover, paradoxically causes the development of many hybrid clones with multiple sets of chromosomes of the unirradiated parent.)

Even more promising is a second technique used by Pontecorvo: one of the parental cell populations is grown for one or two replication cycles in BUdR. Incorporation of this analog renders the chromosomes of these cells subject to breaks upon exposure to visible light (cf. Ref. 205). If the population of cells grown in BUdR is fused with the other, unirradiated parent, the light treatment can be given *after* the formation of hybrids, so that only the chromosomes of the treated parent will undergo breakage and, eventually, elimination. Preliminary experiments indicate that the BUdR + light treatment is as effective as X-irradiation. Pontecorvo rightly points out that this technique, theoretically, offers the possibility of selective incorporation of BUdR and, hence, sensitization and elimination of individual chromosomes according to their time of replication.

Note 15. It must be pointed out that a degree of uncertainty is introduced into linkage determinations by this method by the possibility that the assumption of randomness of the initial rapid loss of human chromosomes may not be entirely correct. This possibility is suspected by several authors, among them G. Marin who stated his opinion in the following terms:

"I believe that in cultured mammalian cells, chromosomes are more often lost in groups, as a consequence of abnormalities in spindle formation. Multipolar spindles are seen quite often in culture. Also, there is some evidence that loss of chromosomes not only tends to occur in groups, but tends not to be random. In other words, if you select for the loss of a specific chromosome, the chance of any other chromosome of the hybrid being lost in association with that one is not the same. This condition would tend to create 'artificial' linkage groups, which are not due to the physical association of genes in the same chromosome. Perhaps these observations do not apply to interspecific hybrids, but this certainly seems to be the case in the hamster intraspecific hybrid which I looked at (G. Marin and J. W. Littlefield [1968]. *J. Virol.*, 2:69; unpublished observations)." (See p. 58: Discussion following paper by Green, Ref. 79; see also Santachiara *et al.*[215])

Note 16. Considerable advances have recently been made in the development of cytochemical techniques allowing the identification of human chromosomes in human × mouse hybrids. Essentially, three new techniques have been developed.

A first technique worked out by Caspersson *et al.*[26] is based on quinicrine-mustard fluorescent staining which reveals a characteristic banding pattern for each pair of homologous human chromosomes. This pattern is different from that of mouse chromosomes, and is conserved in human × mouse hybrid cells.[25, 28] Figure A.1 shows the karyotype of one of the RAG × WI-38 segregated hybrids as visualized by this method.

The description of two other new methods, utilizing the peculiar properties of mouse "constitutive" heterochromatin (which is the source of mouse satellite DNA, clustered in the centromeric regions of mouse chromosomes), will be found in the paper by Chen and Ruddle.[28] These authors convincingly show that in human × mouse hybrids, the mouse chromosomes can be distinguished from the human ones either by *in situ*

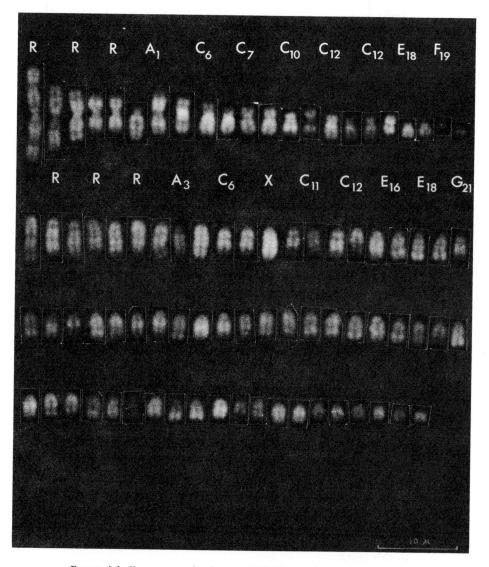

FIGURE A.1. Karyogram of a human (WI-38) × mouse (RAG) hybrid visualized by the quinacrine mustard fluorescence technique. R, RAG bi-armed chromosomes; A1, A3, C6, C7, X, C10, C11, C12, E16, E18, F19, G21 are human chromosomes. Courtesy of T. R. Chen and F. H. Ruddle (unpublished).

annealing of radioactive mouse satellite DNA to intact metaphase chromosomes (a technique developed by Pardue and
Gall[189]) or by a specific staining of mouse centromeric heterochromatin.

Note 17. Paul Weiss[261] expresses the same idea in different terms:
"Many people erroneously imagine that they study differentiation when they study the transformation of a myoblast into
a muscle cell or the development of pigment in a melanoblast,
which is already singletracked and can never do anything except either produce melanin or not produce it; they fail to
distinguish this from strain differentiation, where something is
passed on to the cell progeny which breeds true throughout
subsequent cell generations, even in changed and indifferent
environment."

I may add that undetermined cells can also multiply as such.
An example of this is provided by *in vitro* cultures of a (mouse)
testicular teratoma. Clonal lines of this teratoma have been
shown to retain over numerous cell generations *in vitro* their
multipotentiality, i.e. ability to give rise to a variety of tissues
upon inoculation into appropriate mice.[73, 129]

Note 18. Speaking of ubiquitous (household) enzymes in a Symposium held in 1962, Krooth[143] made the following remark:
"What I use as a rule of thumb, and I would not care to have
to defend this rule, [is] that if an enzyme is present in a white
blood cell or, for that matter, any periferal blood cell, and is
also present in some parenchymatous organ other than the
spleen, the enzyme is ubiquitous. I have not done a large survey
of literature to show that this is truly a universal rule, but it
seems to work. Perhaps, by the end of this conference, I shall be
the proud owner of the world's largest collection of exceptions
to this rule."

It is rather amusing to notice that 8 years later, again referring to this rule, he wrote:[145] "As far as is known, this rule still
seems to hold, but the *converse* does not hold."

Note 19. These notions and terms are obviously closely related to
Foulds' notions of the *total* and the *facultative* genomes, as is
clear from the following quotation from his book:[74] "The

whole of the genome is not in effective use at any one time or place; it can be selectively 'activated.' Selective utilization is now widely accepted as the genetic basis of differentiation in the higher animals . . . a useful distinction may be made here between (1) the *total genome* comprising all the potentially utilizable genetic patterns available in the whole of the genetic material and (2) the *effective genome* comprising the genetic patterns that are in effective use at a particular time and place. In general, 'effective use' will mean being engaged in transcription." (p. 262)

Note 20. It is interesting, in this connection, to reread the following paragraph written by T. H. Morgan[177] in 1934:

"As I have already pointed out, there is an interesting problem concerning the possible interaction between the chromatin of the cells and the protoplasm during development. The visible differentiation of the embryonic cells takes place in the protoplasm. The most common genetic assumption is that the genes remain the same throughout this time. It is, however, conceivable that the genes also are building up more and more, or are changing in some way, as development proceeds in response to that part of the protoplasm in which they come to lie, and that these changes have a reciprocal influence on the protoplasm. It may be objected that this view is incompatible with the evidence that by changing the location of cells, as in grafting experiments and in regeneration the cells may come to differentiate in another direction. But the objection is not so serious as it may appear if the basic constitution of the gene remains always the same, the postulated additions or changes in the genes being of the same order as those that take place in the protoplasm. If the latter can change its differentiation in a new environment without losing its fundamental properties, why may not the genes also? This question is clearly beyond the range of present evidence, but as a possibility it need not be rejected. The answer, for or against such an assumption, will have to wait until evidence can be obtained from experimental investigation." (p. 234)

Note 21. In an earlier discussion[64] of the mechanisms of differentiation and determination, I suggested that differentiation *sensu stricto,* and the oscillations of *in vitro* cultures between the state of overt differentiation and that of undifferentiated stem cells could be based on rather simple factors similar to those operating in the induced biosynthesis of enzymes in bacteria, while the remarkably stable determination of cell types in higher organisms, with its "mutual exclusion feature" may involve a mechanism analogous to that responsible for the inactivation (of most of the genes) of one of the X-chromosomes in mammalian females[157] but operating between "sets" of *nonallelic* genes. The molecular mechanism of this phenomenon remains unknown: different hypotheses of inactivation of one of the Xs and of the maintenance of the activity of the other are lucidly discussed by Lyon in Refs. 158 and 159 to which I call the attention of the readers. The profound analogy (if not similarity) of the phenomena involved in X-chromosome inactivation on the one hand and embryonic determination and differentiation on the other, is clearly apparent in the following quotation from Lyon's most recent paper (from which I have omitted references to the literature):[159]

"In a normal female mammal the two X-chromosomes in each cell become differentiated at an early stage in embryonic development. One retains the euchromatic, genetically active state and the other is late replicating, condensed (forming the sex chromatin body), and inert. In a male the single X remains in the active state, and from individuals with abnormal numbers of X-chromosomes it is clear that the system is such as always to maintain a single X active. The word 'active' here does not of course imply a state of constant transcription. Rather it means a potentiality for gene transcription in response to appropriate inducers or derepressors. Although the term 'inactive X' is now entrenched in the literature, with hindsight 'unresponsive X' might have been more appropriate.

"Once the differentiation has occurred, in embryogeny, each X in a cell normally retains its particular functional state through numerous subsequent mitotic divisions. This can be seen either *in vivo,* through the development of large patches of cells, all with the same X active, or *in vitro* by studies of cultured cells taken from heterozygotes for X-linked marker

genes, such as those for the enzymes G6PD, HGRPT, PGK
and α-galactosidase. Whereas uncloned cultures exhibit the
activity of both the heterozygous alleles, single-cell clones show
that of only one or the other. Thus in considering the mechanism of X inactivation there are two distinct problems: (1) how
is the differentiation brought about, and (2) how is it maintained through subsequent cell divisions and concomitant
chromosome replications?"

Note 22. The reader's attention is called to a recent review by
Davidson[43] wherein (p. 422) some general difficulties of interpretation of results obtained with nondividing heterokaryons
are indicated and some pertinent questions raised.

Concerning the results of Harris' group described up to this
point in Chapter 5, and in particular, the validity of his
speculations on the mechanisms of nuclear reactivation, it may
be pointed out that the entry of cytoplasmic protein into the
enlarging nuclei (to which he ascribes a primary role in nuclear
reactivation) has been observed in a number of cases reviewed
in a recent paper by Merriam[166] where a thought-provoking
discussion of its possible significance in molecular terms will
be found. The original data [concerning the nature of the
proteins moving from the cytoplasm into the nucleus and their
role in the (dissociable) activation of DNA and RNA in brain
and blastula nuclei] presented in this interesting paper suggest
that Harris' interpretations (quoted in Chapter 5) may represent a rather simplified picture of events, and cast additional
doubt on the trivial nature of factors involved in nuclear
synchronization (cf. above Note 10).

Note 23. Nuclear enlargement and induction of DNA and/or
RNA synthesis in nuclei of various (embryonic and adult)
somatic cells have been shown to occur by nuclear transplantation experiments and by fusion of somatic cells into heterokaryons. Since the question raised in the last paragraph quoted
from Harris' book[101] is really that of *reprogramming* of nuclei
of determined or differentiated cells, it is worth noting that
the occurrence of this phenomenon has thus far been established only in the case of transplantation of nuclei of differen-

tiated cells into the cytoplasm of (enucleated) *mature* frog *eggs* and has been shown by Gurdon to be due to substances of presumably nuclear origin (since they appear in the cytoplasm at the moment of the rupture of the germinal vesicle). (For excellent reviews, see Refs. 89, 90, 92, 94, 126, 166.) We shall see later (Note 30) that fibroblast genomes can apparently be reprogrammed (as judged by the production of a liver luxury product) in the cytoplasm of a somatic (neoplastic liver) cell, and that the occurrence of this phenomenon ("induction") seems to be dependent on the dosage of liver cell genes.

Reprogramming of determined or differentiated somatic cells is obviously involved also in the transdetermination observed in *in vivo* cultures of Drosophila imaginal disks,[263] and in metaplasia.

Note 24. In connection with these two points I may mention (1) that a recent paper by Sonnenschein *et al.*[236] describes some evidence for interaction in hybrid cells between the genomes of the parental cells expressed by the level of a household enzyme (thymidine kinase); and (2) that some evidence on complementation between different alleles of the same structural gene has been obtained in hybrids between pairs of human diploid cell lines derived from different patients with galactosemia, i.e. carrying autosomal mutations causing deficiency of galactose-1-phosphate uridyl transferase.[183] Finally, I wish to call the attention of the reader to the beautiful work of Puck's group on the complementation between *glycine⁻* mutants of Chinese hamster cells[130, 131] not mentioned in the text.

Note 25. I was referring here to the observations of Littlefield[152] on several hybrids between "wild type" Syrian hamster cells of line BHK and aminopterin-resistant cells of several clones derived from this line. The aminopterin resistance of the mutants is due to the presence of folate reductase in amounts up to 125 times higher than in the wild type cells, owing apparently to overproduction of the enzyme (rather than to enzyme stabilization). Of the 35 hybrid clones examined, 32 contained levels of folate reductase intermediate between those of the parental lines. Thus, in this case, there is no evidence for interaction between the parental genomes, and Littlefield suggests that the folate reductase overproducing cells may be mutants of the (bacterial) "operator-constitutive" type.

Concerning the question of scarcity of evidence for regulatory genes from hybridization experiments, I wish to call attention of the readers to Littlefield's recent review under the revealing title: "Weak evidence for bacterial type regulation in animal cells,"[153] and to an earlier review by Pardee and Wilson[188] and papers by Eagle and co-workers[53, 54] who also emphasize both the rarity of regulatory effects in animal cells cultivated *in vitro* and the feeble amplitude of substrate-induced changes of enzyme levels as compared with those in bacteria.

The reader will naturally recall that the observations of Harris' group (described in Chapter 5) on the reactivation of nuclei of differentiated cells in heterokaryons with actively growing cells, indicate the operation of mechanisms regulating DNA and *global* RNA synthesis. As we have seen in Chapter 5, tests of heterokaryons for synthesis of defined gene products revealed activation of genes concerned with "household" functions only.

Note 26. It has recently been found by Benda and Davidson[8] that mouse fibroblasts of line 3T3 do not contain any C' fixing material. Hybrids between glial cells and these fibroblasts have been isolated, and are being tested for the presence of S-100. The first results suggest that study of these hybrids will facilitate the interpretation of the results obtained with the glial cell $\times$ clone 1D hybrids.

Note 27. After having studied the fate of S-100 production by glial cell $\times$ fibroblast hybrids (pp. 78ff.), Davidson and Benda[44] have analyzed the behavior in such hybrids of another luxury function of glial cells: the production of large amounts of (*cytoplasmic*) glycerol phosphate dehydrogenase (GPDH), which, *in vitro*, is "inducible" by hydrocortisone (HC) in glial cells, but not in other types of brain cells or in fibroblasts.

Using hybrids between 1*s* or 2*s* glial cells and fibroblasts of lines 3T3 and Cl.1D, Davidson and Benda observed that (1) GPDH does not increase when the hybrids are exposed to HC (i.e. that there is extinction of GPDH inducibility); and (2) there is no correlation between the constant extinction of GPDH inducibility and the more or less pronounced depression of baseline activity of this enzyme in the various hybrids. These facts led the authors to conclude that the induced and noninduced levels of GPDH are controlled independently. Moreover, from comparisons of S-100 production and GPDH baseline activities by the different hybrids, Davidson and Benda infer that all specialized functions characteristic of glial cells are not subject to co-ordinate control. This conclusion is consonant with that which has been arrived at following the study of hepatoma hybrids described in detail in Chapter 6.

Note 28. Using a cytochemical technique for the detection of TAT, Thompson and Gelehrter[246] have studied the induction of this enzyme in heterokaryons formed by the fusion of rat hepatoma cells (line HTC) and BRL-62 cells (isolated by Coon like the BRL cells described above). The noninduced hepatoma cells show slightly positive staining and the latter is much more intense in induced cells. The BRL-62 cells contain no TAT. The heterokaryons examined 24 hours after fusion showed no TAT activity after incubation with Dex. Given the sensitivity of the method and the half-life of the enzyme (or TAT mRNA), the authors conclude that their results are best interpreted as resulting from the almost immediate and total arrest of TAT synthesis at either the transcriptional or translational levels.

It is important to notice that these heterokaryons must have contained *all the chromosomes* of both parents.

Note 29. In the cases of *re-expression* of luxury functions described above, the loss of the chromosomes of the parent not expressing the function studied was spontaneous and apparently random, and the detection or re-expression required biochemical assays. In recent experiments (C. Fougère, F. Ruiz, and B. Ephrussi, *PNAS*, in press), we used Pontecorvo's technique for the induction of directed chromosome losses (irradiation of one of the parents of a cross, as described in Note 14) in the hope of producing melanoma (3460-3) $\times$ (irradiated)

136

fibroblast (Cl.1D) hybrids which would preferentially lose fibroblast chromosomes and resume melanin synthesis, and which would be detectable by visual inspection of hybrid populations. Although the Pontecorvo effect was not obvious in our experiments, they were successful in that the formation of a few melanin producing hybrid clones was indeed observed among a great majority of unpigmented hybrids similar in phenotype and karyotype to those described on p. 73. However (1) karyological analysis of the pigmented hybrids proved them to contain 2s melanoma × 1s fibroblast chromosome complements (with some loss of chromosomes of both species); (2) some unpigmented hybrids were observed within the same populations and these showed karyotypes very similar to the pigmented ones; (3) the derivation of these two types of hybrids from each other within (presumed) clonal populations could not be unequivocably established. Owing to these facts, it is not possible to state at this time whether in the pigment-producing hybrids there is *lack of extinction* or *re-expression* (correlated with loss of some fibroblast chromosomes) of pigment synthesis. The only obvious conclusion from these observations is that there is an effect of the dosage of melanoma genes because melanin-producing hybrid cells have never been found within any of the thousands of colonies of melanoma × fibroblast hybrids with a single set of melanoma genes observed in our laboratory.

This led us to undertake crosses between 2s melanoma (3460-3) cells and unirradiated 1s fibroblasts (Cl.1D). The hybrid colonies obtained in this cross fall into 3 classes: (1) pigmented, (2) unpigmented, (3) mixed—the latter category being the most numerous. The "dosage effect" was thus confirmed (cf. Note 30), but a detailed study of the karyotypic constitution of the hybrids of the three classes and of the mode of formation of mixed colonies is required in order to establish whether the difference between hybrids of classes (1) and (2) is due to the presence *vs.* early loss of some specific, unidentified fibroblast chromosomes producing the hypothetical substance responsible for the extinction of melanogenesis and/or to the effect of the relative dosage of fibroblastic regulatory genes and melanoma genes involved in melanin formation. (Note that there is a systematic difference between the morphologies of the pigment producing and nonproducing cells.)

Note 30. The phenomena of transdetermination and metaplasia (mentioned p. 52), prove that the epigenetic states are subject to rare changes; in other words, they are reversible. Some interesting evidence on this conclusion comes from the study of albumin production by some new recently obtained hepatoma $\times$ fibroblast hybrids. As compared with the results described on p. 84, a completely different set of results has been obtained when 2s hepatoma cells [clone 2s Fu5-5 (1e)] were crossed with 3T3.[194] Five hybrid clones (series 3-2F) have been isolated. Tests for the production of both rat and mouse albumin proved these clones to fall into three classes: (1) one hybrid clone produces *both rat and mouse albumin*; (2) two clones produce *only mouse* albumin; (3) and two clones produce *neither*.

These results are different in two respects from those obtained with the 1s hepatoma hybrids (series 3F); (1) all eight clones of the 3F hybrids secrete rat serum albumin (RSA), while of the five 3-2F clones only one produces RSA; (2) none of the 3F hybrids produce mouse albumin (MSA), while three out of 5 of the 3-2F hybrids do so.

While the reason for the occurrence of the different classes of 3-2F hybrids with respect to the production and kind of albumin remains obscure, the karyotypes of these hybrids make it rather unlikely that they are correlated with the loss of structural genes for albumin. On the other hand, it seems difficult to avoid the conclusion that the production of MSA by three out of five 3-2F hybrids, as compared with the production of *RSA only* by all of the 3F hybrids, is due to the difference in dosage of hepatoma genes (cf. Note 29). If so, it may be concluded that the epigenotype of a fibroblast, *as judged by the production of albumin*, can be *induced* to change to that of a liver (hepatoma) cell under the influence of an overwhelming dosage of hepatoma genes. Moreover, as pointed out by the authors, "the fact that the induction of the mouse fibroblast gene for albumin synthesis occurs in hybrids in which the rat hepatoma gene is not expressed may be interpreted as evidence for the existence of a heritable regulatory element for the activation and maintenance of albumin synthesis, which is separate and independent of the structural gene."

Note 31. In the preceding paragraph I have qualified the phenomenon of extinction as *quasi*-general because the behavior of albumin in hepatoma $\times$ fibroblast hybrids, described on p. 96, makes it questionable whether the term extinction is really appropriate in this case. At the time I gave my lectures, this was the only instance of this sort. During the preparation of my lectures for publication, two reports have appeared which suggest absence of extinction in hybrids between cells one of which exhibits a given differentiated function and the other does not.

(1) Minna *et al.*[171] have studied hybrids between cells of a mouse neuroblastoma and L cells. Among other characteristics, neuroblastoma cells have electrically excitable membranes, while the L cells are electrically passive. The hybrids were found to express "at least a part of the genetic information for neuron differentiation." No data on the other neuroblastoma functions nor on the karyology of these hybrids are given. The latter is particularly regrettable in this case because the electrical measurements were made (as the authors themselves point out) on the *largest* of the hybrid cells. (Cf. Note 29 on the possible relationship between gene dosage and extinction.)

(2) Several authors have published rather conflicting observations on hybrids between antibody producing cells and cells producing no antibody. Hybrids between a mouse plasmacytoma (MOPC 315, which secretes both γG immunoglobulin and lambda light chains) and Cl.1D fibroblasts were found to produce very little, if any, antibody.[192] Total extinction of immunoglobulin production was observed also in hybrids between cells of mouse myeloma MPC 11 (producing both light and heavy chains) and mouse fibroblasts of a clone derived from 3T3.[30] However, absence of extinction of antibody production was observed in (chromosomally nearly complete 1s + 1s) hybrids between cells of a mouse myeloma, producing both γG and free Kappa chains, and cells of a mouse lymphoma pro-

APPENDIX

ducing no immunoglobulin.[172, 173] Of 13 hybrids studied, 11 produce only free Kappa chains, 2 (with the highest chromosome numbers) synthesize both γG and free Kappa chains. An interesting discussion of the possible reasons for the discrepancy between the results of the three studies will be found in Ref. 172. Among the possible explanations offered which would reconcile the three sets of observations, one ascribes the absence of extinction in the myeloma $\times$ lymphoma hybrids to the allotypic specificity of the hypothetical "repressors" causing the extinction of immunoglobulin synthesis.

Note 32. It may be appropriate at this point to remind the reader that this hypothesis, as conceived originally to account for extinction of melanin synthesis in melanoma $\times$ fibroblast hybrids, deals with the properties of cells with different epigenotypes, i.e. with the effects of differentiation and not with its causes (cf. p. 55 and Note 21). In its *simplest form*, it was based on the assumption that the diffusible regulator substance which specifically "represses" the synthesis of dopa-oxidase by the melanoma genome in the hybrids is also responsible for the absence of synthesis of this enzyme in the fibroblasts themselves. The following quotations from a review by Davidson[43] (written in 1969) states the corollaries of this hypothesis: (1) "The conclusion that the repression observed in the hybrids is indicative of the operation of mechanisms of gene regulation is based on the assumption that hybrids between differentiated cells characterized by the same function would express that function; (2) These speculations predict that, in hybrids between differentiated cells which express functions other than those already tested and undifferentiated cells, the differentiated functions would be repressed. . . (3) In addition, it would be expected that, in hybrids between two differentiated cells which express different functions of both parental cells, the differentiated functions of both parental cells would be repressed, since the genome of each cell type would produce a repressor for the differentiated function of the other."

The correctness of (1) has by now been proved by the observations on the similarities of the differentiated phenotypes of 1s and 2s (melanoma, glial, and hepatoma) cells obtained by

140

pairwise fusion of the corresponding 1s cells. The only exception to prediction (1) are observations on hybrids between fibroblasts differing in the rates of collagen synthesis and synthesis vs. absence of synthesis of hyaluronic acid where no interaction between the parental genomes was observed.[80] The significance of this exception is not clear (cf. Ref. 81) and may be trivial, as pointed out by the authors[80] (see also discussion of this case in Ref. 43).

As we have seen in the preceding pages of Chapter 6 and Note 31, prediction (2) is generally fulfilled but there are apparently also some exceptions. These have been listed in Note 31. Among these, two—the absence of total extinction of albumin secretion in hepatoma hybrids (pp. 86 and 88) and the continued production of immunoglobulin by myeloma $\times$ lymphoma hybrids—have in common that we are dealing here not with intracellular enzymes but with secreted products which may be subject to a different type of regulation. This suggestion encounters a difficulty in the observation that another secreted product, pituitary hormone (p. 78) is subject to extinction. This difficulty may however be a spurious one: the pituitary cell $\times$ fibroblast hybrids were apparently of the 1s $\times$ 2s type, so that we may be dealing here with a dosage effect similar to that described in Note 29.

I may add that while I am listing the hybrids between (antibody producing) myeloma cells and lymphoma cells (which produce no antibody) in connection with prediction (2) dealing with hybrids between differentiated and "nondifferentiated" cells, this classification may be erroneous. There is the interesting possibility that the continued production of immunoglobulins by myeloma $\times$ lymphoma hybrids is due to the apparently close embryological relationship (i.e. possibly a partly common type of determination) between these two cell types. Note that this suggestion does not account for the difference between the (1) antibody producing myeloma $\times$ lymphoma hybrids and (2) the myeloma $\times$ fibroblast hybrids which produce no antibody, in terms of the simplest scheme of regulation of differentiated functions (described in the above quotations from Davidson), but that it may be compatible with the two-stage regulation scheme which I referred to in Note 21 and to which I give my preference. Concerning point (3), all that can be said at the present time is that no information on this point has been published as yet.

Note 33. My attention has been called to the possibility that the extinction of a luxury function of a differentiated cell of one type following fusion with a cell of another type may be due to the presence in the latter of an enzyme which destroys something necessary for the differentiation of the former and, hence, that extinction may have nothing to do with the mechanisms regulating normal differentiation. While such a mechanism of extinction cannot be excluded, it appears to me that the presence of such an enzyme characteristic of one cell type and absent in the other, if proved, would be just another expression of their divergent determination or differentiation. Therefore this hypothesis would fall within the realm of those listed in the text: the (hypothetical) enzyme would obviously be an element of the mechanisms regulating differentiation.

Note 34. Experiments on transplantation of nuclei of the frog renal adenocarcinoma into enucleated frog eggs have resulted in the development of a certain number of almost normal tadpoles (for references, see McKinnel *et al.*[164]). The experiments are not flawless: as indicated by the authors themselves, it cannot be excluded that the successful development was not directed by normal nuclei originating from stroma cells rather than from neoplastic cells: the probability of the former being the case appears however to be very low. Therefore, if one assumes that the development of the nearly normal tadpoles was directed by neoplastic nuclei, it follows that (1) the nuclei of the adenocarcinoma cells were still "totipotent" and, hence (2) that the (apparently virus induced) neoplastic state of cells did not involve an irreversible, genetically determined change in their regulatory system; in other words, that an epigenetic change may be responsible for their neoplasticity.

Note 35. Essentially, only one aspect of the cancer problem was discussed in this part of my third lecture. (References to other aspects have been given in Notes 1 and 6.) Even this discussion was very limited both by the lack of lecture time and of time for thoroughly analyzing the numerous data contained in the then as yet unpublished papers by H. Harris, G. Klein, and their co-workers[15, 138, 262] of which the authors kindly sent me the manuscripts shortly before I gave the Carter-Wallace Lectures. Therefore, my presentation was based mostly on the preliminary report by Harris et al.[104] and the text of the lectures is considerably expanded here.

Note 36. Unfortunately, the evaluation of the results reported in this hastily and sensationally written paper was made difficult by the lumping of data concerning the take incidences of A-9 and of the hybrid cells under different test conditions, some of which are not specified in the tabulation of results. Both subcutaneous and intraperitoneal inoculation were made into newborn and adult, irradiated and unirradiated syngeneic and allogenic mice. In spite of the fact that these variables clearly affect the take incidences, the authors draw their conclusions from cumulative take indices, which by comparison, for example, with the results of intraperitoneal inoculation alone, leads to an underestimation of the malignant potential of A-9 (qualified as nonmalignant) and of that of some hybrids. Note that in subsequent publications[15, 138, 262] only results of subcutaneous inoculations are given, while the preliminary report discussed here suggests that the intraperitoneal route results in more takes both of A-9 and of the A-9/Ehrlich hybrids.

Note 37. I may add that in the paper with Defendi, Koprowski, and Yoshida,[51] the conclusion that the malignancy of polyoma transformed × normal T-6 fibroblast hybrids behaves as a dominant trait was accompanied by clear reservations. We pointed out that our interpretation applied only to a cell population "since the test for tumorigenicity selects for the more virulent cells" and that the validity of our conclusion depended on: "(1) The T-6 cells being indeed normal at the time of fusion and (2) at the time of testing no significant losses of the normal alleles contributed by the T-6 parent having occurred in the hybrid cells." After having stated the difficulties of proving points (1) and (2), we said: "With these reservations in mind, it seems reasonable to conclude that the investigated properties of polyoma-transformed cells are expressed in their hybrids with *normal* cells."

Note 38. It may be significant in this respect that successful selection of nonmalignant (contact inhibited) cells from malignant populations *in vitro* (first described by Pollack *et al.*[200] and Rabinowitz and Sachs[206]) is accompanied by marked increase in chromosome number. Moreover, back-reversion to the malignant (noncontact inhibited) phenotype is correlated with a shift to the original stemline number of chromosomes.[112, 201] These observations suggest that the TA_3/fibroblast hybrids selected on the basis of high chromosome numbers[262] may have contained two fibroblast genomes. Thus their low malignancy may reflect a gene dosage effect similar to that described by Murayama-Okabayashi *et al.*[181] and mentioned on p. 107. (For speculations on the mechanism involved, see Refs. 112 and 201.)

Note 39. At this point, I would like to recall the following remark made by Monod and Jacob[174] soon after their discovery of regulatory genes:

"These observations may have some bearings on the problem of the initial event leading to malignancy. Malignant cells have lost sensitivity to the conditions which control multiplication in normal tissues. That the disorder is genetic cannot be doubted. That, following an initial event, mutations within the cellular population are progressively selected, leading towards greater independence, i.e., heightened malignancy, is now quite clear, due in particular to the work of Klein and Klein (1958) [my Ref. 140]. But while the initial event, responsible for setting up the new selective relationships, may of course be a genetic mutation, it might also be brought by the transient action of an agent capable of complexing or inactivating *temporarily* a genetic locus, or a repressor, involved in the control of multiplication. It is clear that a wide variety of agents, from viruses to carcinogenes, might be responsible for such an initial event."

Note 40. The specificity of the surfaces of different embryonic cells was first demonstrated by Holtfreter.[113] Moscona and Moscona[179] gave a similar demonstration for differentiated cells by elegant reaggregation experiments. (Relevant references to further work with Moscona's techniques will be found in Moscona[178] as well as in Pfeiffer *et al.*[196, 197].) In the two papers by the latter authors, numerous examples are cited indicating the relationship between cell surface and expression of luxury functions (see also Refs. 115 and 221). References to studies of the role of cell surfaces in morphogenetic movements will be found in the review by Green and Todaro,[82] while the relationship between cell surface properties, the structure of cell membranes, and growth control (contact inhibition) in normal cells and their transformed (malignant) counterparts are reviewed in two recent symposia.[228, 264] For the role of cell surface properties in the invasiveness of tumor cells, see L. Weiss' book.[252]

Note 41. I would like to call the attention of the readers to the (as usual, stimulating) article by Sonneborn[234] whose views on development of higher organisms are strongly influenced by the observations of his own group and of Tartar[244] on the role of the cortex in the morphogenesis of Protozoa. Following is a brief quotation from this article:

"The inherited determinative pattern of diverse areas in the cortex of the egg could *conceivably* be transmitted by an unbroken chain of cycles of cortical changes each of which depends decisively on the molecular microgeography of the cortex at the start of each cycle. *The required cycle of cortical changes would appear to be minimal, though not inconsiderable, because of the fairly direct route from egg to egg via an early demarkated germ line."* (All italics are mine.)

Extrapolations from Protozoa, which are *noncellular* organisms,[52] to cells of multicellular organisms are obviously dangerous. It is therefore regrettable that experimental evidence supporting cortical inheritance in higher animals is, to this day, limited to observations of a single author.[40] The phenomena of extinction of luxury functions and of their re-expression, as described in Chapter 6 and discussed in Chapter 8, if anything, make cortical inheritance in mammalian somatic cells rather unlikely in my opinion. This, however, is only a guess. If development of higher organisms really is one of the "many things we should like to know," then we must know them "in considerable detail" before we can claim that we understand them at all. (The expressions in quotes are from the epigraph to Chapter 8.)

References

1. Abercrombie, M. (1967). General review of the nature of differentiation. In *Cell Differentiation* (A.V.S. De Reuk and J. Knight, eds.), pp. 3–12. Churchill Ltd., London.

2. Aoki, T., U. Hämmerling, E. de Harven, E. A. Boyse, and L. J. Old (1969). Antigenic structure of cell surfaces. *J. Exp. Med.*, 130:979–1001.

3. Attardi, G., H. Parnas, and B. Attardi (1970). Pattern of RNA synthesis in duck erythrocytes in relationship to the stage of cell differentiation. *Exp. Cell Res.*, 62:11–31.

4. Barski, G. and F. Cornefert (1962). Characteristics of "Hybrid"-type clonal cell lines obtained from mixed culture *in vitro*. *J. Nat. Cancer Inst.*, 28:801–821.

5. ———, S. Sorieul, and F. Cornefert (1960). Production dans des cultures *in vitro* de deux souches cellulaires en association, de cellules de caractère "hybride." *C. R. Acad. Sci. (Paris)*, 251:1825–1827.

6. ———, S. Sorieul, and F. Cornefert (1961). "Hybrid" type cells in combined cultures of two different mammalian cell strains. *J. Nat. Cancer Inst.*, 26:1269–1291.

7. Beisson J. and T. M. Sonneborn (1965). Cytoplasmic inheritance of the organization of the cell cortex in *Paramecium aurelia*. *Proc. Nat. Acad. Sci., USA*, 35:275–282.

8. Benda, P. and R. L. Davidson (1971). Regulation of specific functions of glial cells in somatic hybrids. I. Control of S-100 protein. *J. Cell. Physiology* , 78:209-216.

9. ———, J. Lightbody, G. Sato, L. Levine, and W. Sweet (1968). Differentiated rat glial cell strain in tissue culture. *Science*, 161:370–371.

10. Bertolotti, R. and M. Weiss (1972). Expression of differentiated functions in hepatoma cell hybrids. II. Aldolase, *J. Cell. Physiology*, 79:211-224.

11. Bolund, L., N. R. Ringertz, and H. Harris (1969). Changes in the cytochemical properties of erythrocyte nuclei reactivated by cell fusion. *J. Cell Sci.*, 4:71–87.

12. Bonner, J. T. (1960). The unsolved problem of develop-

ment: an appraisal of where we stand. *Amer. Scientist*, 48: 514–527.

13. Boone, C. M. and F. H. Ruddle (1969). Interspecific hybridization between human and mouse somatic cells: enzyme and linkage studies. *Biochem. Genet.*, 3:119–136.

14. Braun, A. C. (1969). *The cancer problem: a critical analysis and modern synthesis.* Columbia Univ. Press, New York.

15. Bregula, U., G. Klein, and H. Harris (1971). The analysis of malignancy by cell fusion. II. Hybrids between Ehrlich cells and normal diploid cells. *J. Cell Sci.*, 8:673–680.

16. Britten, J. and E. H. Davidson (1969). Gene regulation for higher cells: a theory. *Science*, 165:349–357.

17. Brown, D. D. and I. B. Dawid (1969). Developmental genetics. *Ann. Rev. Genetics*, 3:127–154.

18. Bruns, G. P. and I. M. London (1965). The effect of hemin on the synthesis of globin. *Biochem. Biophys. Res. Comm.*, 18:236–242.

19. Burger, M. M. (1971). Structural changes of the surface membrane after viral transformation: Forssman antigen belongs to the exposed and not the induced antigens. *Nature*, 231:125–126.

20. ——— and K. D. Noonan (1970). Restoration of normal growth by covering of agglutinin sites on tumour cell surface. *Nature*, 228:512–515.

21. ———, K. D. Noonan, Jr. R. Sheppard, T. O. Fox, and A. J. Levine (1971). Requirement for the formation of a structural surface change after viral infection and the significance of this change for growth control. In *2nd Lepetit Colloquium on the Biology of Oncogenic Viruses*, pp. 258–267. North Holland, Amsterdam.

22. Burnett, M. F. (1967). The impact of ideas on immunology. *Cold Spring Harbor Symp. Quant. Biol.*, 32:1–8.

23. Cahn, R. D. and M. B. Cahn (1966). Heritability of cellular differentiation: clonal growth and expression of differentiation in retinal pigment cells *in vitro*. *Proc. Nat. Acad. Sci., USA*, 55:106–114.

24. Calissano, P. and G. Rusca (1969). On the "heterogeneity" of the beef brain S-100, a specific nervous tissue protein. In

148

2nd Intern. Meeting, Intern. Soc. Neurochem. (R. Paoletti, ed.), p. 115. Tamburin Editors, Milan.

25. Caspersson, T., L. Zech, H. Harris, F. Wiener, and G. Klein (1971). Identification of human chromosomes in a mouse/human hybrid by fluorescence techniques. *Exp. Cell Res.*, 65:475–478.

26. ———, L. Zech, C. Johansson, and E. J. Modest (1970). Identification of human chromosomes by DNA-binding fluorescent agents. *Chromosoma*, 30:215–227.

27. Cassingena, R., C. Chany, M. Vignal, H. Suarez, S. Estrade, and P. Lazar (1971). Use of monkey-mouse hybrid cells for the study of the cellular regulation of interferon production and action. *Proc. Nat. Acad. Sci., USA*, 68:580–584.

28. Chen, T. R. and F. H. Ruddle (1971). Karyotype analysis utilizing differentially stained constituitive heterochromatin of human and murine chromosomes. *Chromosoma*, 34:51–72.

29. Christen, Ph., U. Rensing, A. Schmid, and F. Leuthardt (1966). Multiple Formen der Aldolase in Organextrakten der Ratte. *Helvetica Chimica Acta*, 49:1872–1875.

30. Coffino, P., B. Knowles, S. G. Nathenson, and M. D. Scharff (1971). Suppression of immunoglobulin synthesis by cell hybridization. *Nature New Biol.*, 231:87–90.

31. Cook, P. R. (1970). Species specificity of an enzyme determined by an erythrocyte nucleus in an interspecific hybrid cell. *J. Cell Sci.*, 7:1–3.

32. Coon, H. G. (1966). Clonal stability and phenotypic expression of chick cartilage cells *in vitro*. *Proc. Nat. Acad. Sci., USA*, 55:66–73.

33. ——— (1968). Clonal culture of differentiated liver cells. *J. Cell Biol.*, 39:29A.

34. ——— (1969). Clonal culture of differentiated cells from mammals: rat liver cell culture. *Carnegie Institution of Washington Yearbook*, 67:419–427.

35. ———, and M. C. Weiss (1969). A quantitative comparison of formation of spontaneous and virus produced viable hybrids. *Proc. Nat. Acad. Sci., USA*, 62:852–859.

36. ———, and M. C. Weiss (1969). Sendaï produced somatic cell hybrids between L cell strains and between liver and L

149

cells. In *Heterospecific Genome Interaction*, The Wistar Institute Symposium Monog. No. 9 (V. Defendi, ed.) pp. 83–96. The Wistar Institute Press, Philadelphia.

37. Crick, F. (1970). Molecular biology in the year 2000. *Nature*, 228:613–615.

38. Croce, C. M., W. Sawicki, D. Kritchevsky, and H. Koprowski (1971). Induction of homokaryocyte, heterokaryocyte, and hybrid formation by lysolecthin. *Exp. Cell Res.*, 67:427–435.

39. Crowle, A. J. (1961). *Immunodiffusion*. Academic Press, New York.

40. Curtis, A.S.G. (1965). Cortical inheritance in the amphibian *Xenopus lævis*: preliminary results. *Arch. de Biol.*, 76:523–546.

41. Davidson, R. L. (1969). Interactions between genomes in somatic cell hybrids: studies on the regulation of differentiation. In *Heterospecific Genome Interaction*, The Wistar Institute Symposium Monog. No. 9 (V. Defendi, ed.) pp. 97–112. The Wistar Institute Press, Philadelphia.

42. ———— (1969). Regulation of melanin synthesis in mammalian cells, as studied by somatic hybridization. III. A method of increasing the frequency of cell fusion. *Exp. Cell Res.*, 55:424–426.

43. ———— (1971). Regulation of gene expression in somatic cell hybrids: a review. *In Vitro*, 6:411–426.

44. ————, and P. Benda (1970). Regulation of specific functions of glial cells in somatic hybrids. II. Control of inducibility of glycerol-3-phosphate dehydrogenase. *Proc. Nat. Acad. Sci., USA,* 67:1870–1877.

45. ————, and B. Ephrussi (1965). A selective system for the isolation of hybrids between L cells and normal cells. *Nature*, 205:1170.

46. ————, and B. Ephrussi (1970). Factors influencing the "effective mating rate" of mammalian cells. *Exp. Cell Res.*, 61:222–226.

47. ———— , B. Ephrussi, and K. Yamamoto (1966). Regulation of pigment synthesis in mammalian cells, as studied by somatic hybridization. *Proc. Nat. Acad. Sci., USA,* 56:1437–1440.

48. ————, B. Ephrussi, and K. Yamamoto (1968). Regulation

of melanin synthesis in mammalian cells, as studied by somatic hybridization. I. Evidence for negative control. *J. Cell. Physiology,* 72:115–127.

49. ————, and K. Yamamoto (1968). Regulation of melanin synthesis in mammalian cells, as studied by somatic hybridization. II. The level of regulation of 3, 4-dihydroxyphenilalanine oxidase. *Proc. Nat. Acad. Sci., USA,* 60:894–901.

50. Defendi, V., B. Ephrussi, and H. Koprowski (1964). Expression of polyoma induced cellular antigen in hybrid cells. *Nature,* 203: 495–496.

51. ————, B. Ephrussi, H. Koprowski, and M. C. Yoshida (1967). Properties of hybrids between polyoma-transformed and normal mouse cells. *Proc. Nat. Acad. Sci., USA,* 57:299–305.

52. Dobell, C. C. (1911). The principles of protistology. *Arch. J. Protistenk,* 23:269–310.

53. Eagle, H. (1965). Metabolic controls in cultured mammalian cells. *Science,* 148:42–51.

54. ————, C. L. Washington, and M. Levy (1965). End product control of amino acid synthesis by cultured human cells. *J. Biol. Chem.,* 240:3944–3950.

55. Einstein, A. (1945). Reply to criticisms. In *Albert Einstein: Philosopher-Scientist* (P. A. Schlipp, ed.), p. 688. The Library of Living Philosophers, Inc., Evanston, Ill.

56. Eliceiri, G. L. and H. Green (1969). Ribosomal RNA synthesis in human-mouse hybrid cells. *J. Mol. Biol.,* 41:253–260.

57. Engel, E., J. Empson, and H. Harris (1971). Isolation and karyotypic characterization of segregants of intraspecific hybrid somatic cells. *Exp. Cell Res.,* 68:231–234.

58. ————, B. J. McGee, and H. Harris (1969). Recombination and segregation in somatic cell hybrids. *Nature,* 223:152–155.

59. ————, B. J. McGee, and H. Harris (1969). Cytogenetic and nuclear studies on A-9 and B82 cells fused together by Sendaï virus: the early phase. *J. Cell Sci.,* 5:93–120.

60. Ephrussi, B. (1964). Chromosomal markers. In *Somatic Cell Genetics* (R. S. Krooth, ed.), pp. 253-273. University of Michigan Press, Ann Arbor, Michigan.

61. Ephrussi, B. (1965). Hybridization of somatic cells and phenotypic expression. In *Developmental and Metabolic Control Mechanisms and Neoplasia* (The University of Texas M.D. Anderson Hospital and Tumor Institute at Houston, 19th Annual Symp. on Fundamental Cancer Research, 1965), pp. 486–503. Williams and Wilkins Co., Baltimore, Maryland.

62. ——— (1966). Interspecific somatic hybrids. *In Vitro*, 2: 40–45.

63. ——— (1966). Hybridization of somatic cells: introduction. In *Genetic Variation in Somatic Cells* (Symposium on the Mutation Process), pp. 55–60. Academia, Prague.

64. ——— (1970). Somatic hybridization as a tool for the study of normal and abnormal growth and differentiation. In *Genetic Concepts and Neoplasia* (The University of Texas M.D. Anderson Hospital and Tumor Institute at Houston, 23rd Annual Symp. on Fundamental Cancer Research, 1969), pp. 9–28. Williams and Wilkins Co., Baltimore, Maryland.

65. ———, R. L. Davidson and M. C. Weiss (1969). Malignancy of somatic cell hybrids. *Nature*, 224:1314–1315.

66. ———, L. J. Scaletta, M. A. Stenchever, and M. C. Yoshida (1964). Hybridization of somatic cells *in vitro*. In *Symp. Intern. Soc. for Cell Biol. on Cytogenetics of Cells in Culture*, Vol. 3:13–25. Academic Press, New York.

67. ———, and S. Sorieul (1962). Nouvelles observations sur l'hybridation *"in vitro"* de cellules de souris. *C. R. Acad. Sci. (Paris)*, 254:181–182.

68. ———, and S. Sorieul (1962). Mating of somatic cells *in vitro*. In *Approaches to the Genetic Analysis of Mammalian Cells*, pp. 81–97. University of Michigan Press, Ann Arbor, Mich.

69. ———, M. A. Stenchever, and L. J. Scaletta (1964). Hybridization as a tool for cell genetics. In *2nd Inter. Conf. on Congen. Malformations*, pp. 85–93. The Intern. Med. Cong. Ltd., New York.

70. ———, and M. C. Weiss (1965). Interspecific hybridization of somatic cells. *Proc. Nat. Acad. Sci., USA*, 53:1040–1042.

71. ———, and M. Weiss (1967). Regulation of the cell cycle

in mammalian cells: Inferences and speculations based on observations of interspecific somatic hybrids. In *Control Mechanisms in Developmental Processes* (M. Locke, ed.), pp. 136–169. Academic Press, New York.

72. ———, and M. Weiss (1969). Hybrid somatic cells. *Sci. Amer.*, 220:26–35.

73. Finch, B. W. and B. Ephrussi (1967). Retention of multiple developmental potentialities by cells of a mouse testicular teratocarcinoma during prolonged culture *in vitro* and their extinction upon hybridization with cells of permanent lines. *Proc. Nat. Acad. Sci., USA,* 57:615–621.

74. Foulds, L. (1969). *Neoplastic Development*, Vol. 1. Academic Press, New York.

75. Frye, L. D. and M. Edidin (1970). The rapid intermixing of cell surface antigens after formation of mouse-human heterokaryons. *J. Cell Sci.*, 7:319–335.

76. Ganshow, R. (1966). Glucuronidase gene expression in somatic hybrids. *Science*, 153:84–85.

77. Gershon, D. and L. Sachs (1963). Properties of a somatic hybrid between mouse cells with different genotypes. *Nature,* 198:912–913.

78. Graham, C. F. (1969). The fusion of cells with one- and two-cell mouse embryos. In *Heterospecific Genome Interaction*, The Wistar Institute Symp. Monog. No. 9 (V. Defendi, ed.) pp. 19–33. The Wistar Institute Press, Philadelphia.

79. Green, H. (1969). Prospects for the chromosomal localization of human genes in human-mouse somatic cell hybrids. In *Heterospecific Genome Interaction*, The Wistar Institute Symp. Monog. No. 9 (V. Defendi, ed.) pp. 51–58. The Wistar Institute Press, Philadelphia.

80. ———, B. Ephrussi, D. Hamerman, and M. C. Yoshida (1966). Synthesis of collagen and hyaluronic acid by fibroblast hybrids. *Proc. Nat. Acad. Sci., USA,* 55:41–44.

81. ———, B. Goldberg, and G. J. Todaro (1966). Differentiated cell types and the regulation of collagen synthesis. *Nature*, 212:631–633.

82. ——— and G. J. Todaro (1967). The mammalian cell as differentiated microorganism. *Ann. Rev. Microb.*, 21:573–600.

83. Green, H., R. Wang, C. Basilico, R. Pollack, T. Kusano, and J. Salas (1971). Mammalian somatic cell hybrids and their susceptibility to viral infection. *Fed. Proc.*, 30:930–934.

84. Grobstein, C. (1959). Differentiation of vertebrate cells. In *The Cell* (J. Brachet and A. E. Mirsky, eds.), Vol. 1, pp. 437–496. Academic Press, New York.

85. ——— (1966). What we do not know about differentiation. *Amer. Zool.*, 6:89–95.

86. ——— (1967). General discussion. In *Cell Differentiation* (A.V.S. De Reuk and J. Knight, eds.), p. 243. J. A. Churchill Ltd., London.

87. Gross, P. R. (1968). Biochemistry of differentiation. *Ann. Rev. Biochem.*, 37:631–660.

88. Gurdon, J. B. (1962). The developmental capacity of nuclei taken from intestinal epithelium cells of feeding tadpoles. *J. Embryol. Exp. Morphol.*, 10:622–640.

89. ——— (1970). Nuclear transplantation and the control of gene activity in animal development. *Proc. Royal Soc. London B*, 176:303–314.

90. ——— and C. F. Graham (1967). Nuclear changes during cell differentiation. *Sci. Prog., Oxford*, 55:259–277.

91. ——— and V. Uehlinger (1966). "Fertile" intestine nuclei. *Nature*, 210:1240–1241.

92. ——— and H. R. Woodland (1968). The cytoplasmic control of nuclear activity in animal development. *Biol. Rev.*, 43:233–267.

93. ——— and H. R. Woodland (1970). On the long-term control of nuclear activity during cell differentiation. In *Current Topics in Developmental Biology* (Moscona and Monroy, eds.), Vol. 5 pp. 39–70. Academic Press, New York.

94. Hadorn, E. (1966). Dynamics of determination. In *Major Problems in Developmental Biology* (M. Locke, ed.), pp. 85–104, Academic Press, New York.

95. Haldane, J.B.S. (1952). Variation. *New Biol.*, 12:15.

96. Hämmerling, J. (1953). Nucleo-cytoplasmic relationship in the development of *Acetabularia*. *Int. Rev. Cytol.*, 2:475–498.

97. ——— (1963). Nucleo-cytoplasmic interactions in *Acetabularia* and other cells. *Ann. Rev. Pl. Physiol.*, 14:65–92.

98. Harris, H. (1965). Behavior of differentiated nuclei in heterokaryons of animal cells from different species. *Nature,* 206:583–588.

99. ——— (1967). The reactivation of the red cell nucleus. *J. Cell Sci.,* 2:23–32.

100. ——— (1968). *Nucleus and Cytoplasm.* Clarendon Press, Oxford.

101. ——— (1970). *Cell Fusion.* Harvard Univ. Press, Cambridge, Mass.

102. ——— and P. R. Cook (1969). Synthesis of an enzyme determined by an erythrocyte nucleus in a hybrid cell. *J. Cell Sci.,* 5:121–134.

103. ——— and G. Klein (1969). Malignancy of somatic cell hybrids. *Nature,* 224:1315–1316.

104. ———, O. J. Miller, G. Klein, P. Worst, and T. Tachibana (1969). Suppression of malignancy by cell fusion. *Nature,* 223:363–368.

105. ———, E. Sidebottom, D. M. Grace, and M. E. Bramwell (1969). The expression of genetic information: a study with hybrid animal cells. *J. Cell Sci.,* 4:499–525.

106. ——— and J. F. Watkins (1965). Hybrid cells derived from mouse and man: artificial heterokaryons of mammalian cells from different species. *Nature,* 205:640–646.

107. ———, J. F. Watkins, C. E. Ford, and G. I. Schoefl (1966). Artificial heterokaryons of animal cells from different species. *J. Cell Sci.,* 1:1–30.

108. Harris, M. (1964). *Cell Culture and Somatic Variation.* Holt, Rinehart, and Winston, Inc., New York.

109. Hershey, A. D. (1970). Genes and hereditary characteristics. Ann. Rep. of the Director, Genetics Research Unit. *Carnegie Institution Year Book 1968,* pp. 655–668.

110. ——— (1970). Genes and hereditary characteristics. *Nature,* 226:697–700.

111. Hertwig, O. (1875). Beiträge zur Kentniss der Bildung, Befruchtung und Teilung des tierischen Eies. *Morphologisches Jahrbuch,* pp. 347–434.

112. Hitotsumachi, S., Z. Rabinovitz, and L. Sachs (1971). Chromosomal control of reversion in transformed cells. *Nature,* 231:511–514.

113. Holtfreter, J. (1939). Gevebeaffinität, ein Mittel der embryonalen Formbildung. *Arch. Exp. Zellforsch. Gevebezücht.* 23:169–209.

114. Holtzer, H. and J. Abbott (1968). Oscillations of the chondrogenic phenotype *in vitro*. In *The Stability of the Differentiated State* (H. Ursprung, ed.), pp. 1–16. Springer-Verlag, Berlin, Heidelberg.

115. ———, R. Bischoff, and S. Chacko (1969). Activities of the cell surface during myogenesis and chondrogenesis. In *Cellular Recognition* (T. Sneith and R. A. Good, eds.), Meredith Corporation, New York.

116. Hu, F. and P. F. Lesney (1964). The isolation and cytology of two pigment cell strains from B16 mouse melanoma. *Cancer Res.*, 24:1634–1643.

117. Hunter, R. L. and C. L. Markert (1957). Histochemical demonstration of enzymes separated by zone electrophoresis in starch gels. *Science*, 125:1294–1295.

118. Hyden, H. and B. McEwen (1964). A glial protein specific for the nervous system. *Proc. Nat. Acad. Sci., USA,* 55: 354–358.

119. Jacob, F. and J. Monod (1961). Genetic regulatory mechanisms in the synthesis of proteins. *J. Mol. Biol.,* 3:318–356.

120. ——— and J. Monod (1970). Introduction. In *The Lactose Operon* (J. R. Bekwith and D. Zisper, eds.), pp. 1–4. Cold Spring Harbor Laboratory, New York.

121. Jakob, H. and F. Ruiz (1970). Preferential loss of kangaroo chromosomes in hybrids between Chinese hamster and kangaroo-rat somatic cells. *Exp. Cell Res.*, 62:310–314.

122. Jami, J., S. Grandchamp, and B. Ephrussi (1971). Sur le comportement caryologique des hybrides cellulaires homme × souris. *C. R. Acad. Sc. (Paris)*, 272:323–326.

123. ——— and S. Grandchamp (1971). Karyological properties of human × mouse hybrids. *Proc. Nat. Acad. Sci., USA*, 68: 3097–3101.

124. Johnson, R. T. and H. Harris (1969). DNA Synthesis and mitosis in fused cells. III. HeLa-Ehrlich heterokaryons. *J. Cell Sci.,* 5:645–697.

125. ——— and P. N. Rao (1970). Mammalian cell fusion. II.

Induction of premature chromosome condensation in inter-phase nuclei. *Nature*, 226:717–722.

126. ——— and P. N. Rao (1971). Nucleo-cytoplasmic interactions in the achievement of nuclear synchrony in DNA synthesis and mitosis in multinucleate cells. *Biol. Rev.*, 46: 97–155.

127. ———, P. N. Rao, and H. D. Hughes (1970). Mammalian cell fusion. III. A HeLa cell inducer of premature chromosome condensation active in cells from a variety of species. *J. Cell. Physiology*, 76:151–157.

128. Kabat, D. and G. Attardi (1967). Synthesis of chicken hemoglobins during erythrocyte differentiation. *Biochem. Bioph. Acta*, 138:382-399.

129. Kahan, B. W. and B. Ephrussi (1970). Developmental potentialities of clonal *in vitro* cultures of mouse testicular teratoma. *J. Nat. Cancer Inst.*, 44:1015–1036.

130. Kao, F. T., L. Chasin, and T. T. Puck (1969). Genetics of somatic mammalian cells, X. Complementation analysis of glycine-requiring mutants. *Proc. Nat. Acad. Sci., USA*, 64: 1284–1291.

131. ———, R. T. Johnson, and T. T. Puck (1968). Complementation analysis on virus-fused Chinese hamster cells with nutritional markers. *Science*, 144:312–314.

132. ———, and T. Puck (1970). Genetics of somatic mammalian cells: linkage studies with human-Chinese hamster cell hybrids. *Nature*, 228:329–332.

133. Kato, H. and A. A. Sandberg (1968). Chromosome pulverization in Chinese hamster cells induced by Sendaï virus. *J. Nat. Cancer Inst.*, 41:1117–1123.

134. ———, and A. A. Sandberg (1968). Cellular phase of chromosome pulverization induced by Sendaï virus. *J. Nat. Cancer Inst.*, 41:1125–1131.

135. Kessler, D., L. Levine, and F. Fasman (1968). Some conformational and immunological properties of a bovine brain acidic protein (S-100). *Biochem.*, 7:758–764.

136. Klebe, R. J., T. Chen, and F. H. Ruddle (1970). Mapping of a human genetic regulator element by somatic cell genetic analysis. *Proc. Nat. Acad. Sci., USA,* 66:1220–1227.

137. Klebe, R. J., T. R. Chen, and F. H. Ruddle (1970). Controlled production of proliferating somatic cell hybrids. *J. Cell Biol.*, 45:74–82.

138. Klein, G., U. Bregula, F. Weiner, and H. Harris (1971). The analysis of malignancy by cell fusion. I. Hybrids between tumour cells and L cell derivatives. *J. Cell Sci.*, 8:659–672.

139. ———, U. Gars, and H. Harris (1970). Isoantigen expression in hybrid mouse cells. *Exp. Cell Res.*, 62:149–160.

140. ——— and E. Klein (1958). Histocompatibility changes in tumors. *J. Cell. Comp. Physiol.*, 52:125–168.

141. Konigsberg, I. R. (1963). Clonal analysis of myogenesis. *Science*, 140:1273–1284.

142. Koprowski, H., F. C. Jensen, and Z. Steplewski (1967). Activation of production of infectious tumor virus SV 40 in heterokaryon cultures. *Proc. Nat. Acad. Sci.*, 58:127–133.

143. Krooth, R. S. (1964). Study of galactosemia, acatalesemia, and other human metabolic mutants in cell culture. In *Somatic Cell Genetics*, 4th Macy Conf. on Genetics (R. S. Krooth, ed.), p. 168. Univ. of Michigan Press, Ann Arbor, Mich.

144. ———, G. A. Darlington, and A. A. Velazquez (1968). The genetics of cultured mammalian cells. *Ann. Rev. Genetics*, 2:141–164.

145. ——— and E. K. Sell (1970). The action of Mendelian genes in human diploid cell strains. *J. Cell. Physiology*, 76:311–330.

146. Kusano, T., C. Long, and H. Green (1971). A new reduced human-mouse somatic cell hybrid containing the human gene for adenine phosphoribosyltransferase. *Proc. Nat. Acad. Sci., USA*, 68:82–86.

147. Laskey, R. A. and J. B. Gurdon (1970). Genetic content of adult somatic cells tested by nuclear transplantation from cultured cells. *Nature*, 228:1332–1334.

148. Levine, L. and B. W. Moore (1965). Structural relatedness of a vertebral brain acidic protein as measured immunochemically. *Neurosci. Res. Progress Bull.*, 3:18–22.

149. Lin, E.C.C. and W. E. Knox (1957). Adaptation of the rat

liver tyrosine-α-ketoglutarate transaminase. *Bioch. Bioph. Acta,* 26:85–88.

150. Littlefield, J. (1964). Selection of hybrids from mating of fibroblasts *in vitro* and their presumed recombinants. *Science,* 145:709–710.

151. ——— (1966). The use of drug resistant markers to study the hybridization of mouse fibroblasts. *Exp. Cell Res.,* 41: 190–196.

152. ——— (1969). Hybridization of hamster cells with high and low folate reductase activity. *Proc. Nat. Acad. Sci., USA,* 62: 88–95.

153. ——— (1970). Weak evidence for bacterial-type regulation in animal cells. In *Genetics Concepts and Neoplasia* (The University of Texas M. D. Anderson Hospital and Tumor Institute at Houston, 23rd Annual Symposium on Fundamental Cancer Research, 1969), pp. 439–451. The Williams and Wilkins Co., Baltimore, Maryland.

154. London Conference of the normal human karyotype (1963). *Cytogenetics,* 2:264–268.

155. Lucy, J. A. (1970). The fusion of biological membranes. *Nature,* 227:815–817.

156. Luria, S. E. (1966). Macromolecular metabolism. *Suppl. J. Gen. Physiol.,* 49:330.

157. Lyon, M. (1961). Gene action in the X-chromosome of the mouse *(Mus musculus L.). Nature,* 190:372–373.

158. ——— (1968). Chromosomal and subchromosomal inactivation. *Ann. Rev. Genet.,* 2:31–52.

159. ——— (1971). Possible mechanisms of X-chromosome inactivation. *Nature New Biol.,* 232:229–232.

160. Markert, C. L. (1968). Neoplasia: a disease of cell differentiation. *Cancer Res.,* 28:1908–1914.

161. ——— and F. Miller (1959). Multiple forms of enzymes: tissue, ontogenetic, and species specific patterns. *Proc. Nat. Acad. Sci., USA,* 45:753–763.

162. Matsuya, Y. and H. Green (1969). Hybrids between the established human line D98 (presumptive HELA) and 3T3. *Science,* 163:697–698.

163. ———, H. Green, and C. Basilico (1968). Properties and

uses of human-mouse hybrid cell lines. *Nature*, 220:1199–1202.

164. McKinnell, R. G., B. A. Deggins, and D. D. Labat (1969). Transplantation of pluripotential nuclei from triploid frog tumors. *Science*, 165:394–396.

165. Meera Khan, P., A. Westerveld, K. H. Grzeschik, B. F. Deys, O. M. Garson, and M. Siniscalco (1971). X-linkage of human phosphoglycerate kinase confirmed in man-mouse and man-Chinese hamster somatic cell hybrids. *Amer. J. Human Genetics*, 23:614–23.

166. Merriam, R. W. (1969). Movement of cytoplasmic proteins into nuclei induced to enlarge and initiate DNA or RNA synthesis. *J. Cell Sci.*, 5:333–349.

167. Migeon, B. R. and C. S. Miller (1968). Human-mouse somatic cell hybrids with single human chromosome (group E): link with thymidine kinase activity. *Science*, 162:1005–1006.

168. ———, S. W. Smith, and C. L. Leddy (1969). The nature of thymidine kinase in the human-mouse hybrid cell. *Biochem. Genetics*, 3:583–590.

169. Miggiano, V., M. Nabholz, and W. Bodmer (1969). Hybrids between human leukocytes and a mouse cell line: production and characterization. In *Heterospecific Genome Interaction*, The Wistar Institute Symp. Monog. No. 9 (V. Defendi, ed.), pp. 61-76. The Wistar Institute Press, Philadelphia.

170. Miller, O. J., P. R. Cook, P. Meera Khan, S. Shin, and M. Siniscalco (1971). Mitotic separation of two human X-linked genes in man-mouse somatic cell hybrids. *Proc. Nat. Acad. Sci., USA*, 68:116–120.

171. Minna, J., Ph. Nelson, J. Peacock, D. Glazer, and M. Nirenberg (1971). Genes for neuronal properties expressed in neuroblastoma $\times$ L cell hybrids. *Proc. Nat. Acad. Sci., USA*, 68:234–239.

172. Mohit, B. (1971). Gamma G immunoglobulin and free Kappa chain synthesis in different clones of a hybrid cell line. *Proc. Nat. Acad. Sci., USA*, 68:3045–3048.

173. ——— and K. Fan (1971). Hybrid cell line from immuno-

globulin-producing mouse myeloma and a nonproducing mouse lymphoma. *Science*, 171:75–77.

174. Monod, J. and F. Jacob (1961). Teleonomic mechanisms in cellular metabolism, growth, and differentiation. *Cold Spring Harbor Symp. Quant. Biol.*, 26:389–401.

175. Moore, B. W. (1965). A soluble protein characteristic of the nervous system. *Bioch. Biophys. Res. Com.*, 19:739–744.

176. Moore, G. (1964). *In vitro* cultures of a pigmented hamster melanoma. *Exp. Cell Res.*, 36:422–423.

177. Morgan, T. H. (1934). *Embryology and genetics.* Columbia Univ. Press, New York.

178. Moscona, A. A. (1971). Embryonic and neoplastic cell surfaces: availability of receptors for Concanavalin A and wheat germ agglutinin. *Science*, 171:905–907.

179. ——— and H. Moscona (1952). The dissociation and aggregation of cells from organ rudiments of the early chick embryo. *J. Anat.*, 88:287–301.

180. Murayama, F. and Y. Okada (1970). Appearance, characteristics and malignancy of somatic hybrid cells between L and Ehrlich ascites tumor cells formed by artificial fusion with UV-HVJ. *Biken's J.*, 13:11–23.

181. Murayama-Okabayashi, F., Y. Okada, and T. Tachibana (1971). A series of hybrid cells containing different ratios of parental chromosomes formed by two steps of artificial fusion. *Proc. Nat. Acad. Sci., USA*, 68:38–42.

182. Nabholz, M., V. Miggiano, and W. Bodmer (1969). Genetic analysis with human-mouse somatic cell hybrids. *Nature*, 223:358–363.

183. Nadler, H. L., C. M. Chacko, and M. Rachmeler (1970). Intrallelic complementation in hybrid cells derived from human diploid cells deficient in galactose-1-phosphate uridyl transferase activity. *Proc. Nat. Acad. Sci., USA*, 67:976–982.

184. Okada, Y. (1958). The fusion of Ehrlich's tumor cells caused by HVJ virus *in vitro. Biken's J.*, 1:103–110.

185. ——— (1962). Analysis of giant polynuclear cell formation caused by HVJ virus from Ehrlich's ascites tumor cells. I. Microscopic observation of giant polynuclear cell formation. *Exp. Cell Res.*, 26:98–107.

186. Okada, Y. (1969). Factors in fusion of cells by HVJ. In *Current Topics in Microbiology*, 48:102–128. Springer Verlag, Berlin, Heidelberg.

187. —— and J. Tadokoro (1962). Analysis of giant polynuclear cell formation caused by HVJ virus from Ehrlich's ascites tumor cells. *Epx. Cell Res.*, 26:108–118.

188. Pardee, A. B. and A. C. Wilson (1963). Control of enzyme activity in higher animals. *Cancer Res.*, 23:1483–1490.

189. Pardue, M. L. and J. G. Gall (1970). Chromosomal localization of mouse satellite DNA. *Science*, 168:1356–1358.

190. Penhoet, E., M. Kochman, R. Valentine, and W. J. Rutter (1967). The subunit structure of mammalian fructose diphosphate aldolase. *Biochemistry*, 6:2940–2949.

191. ——, T. Rajkjmar, and W. J. Rutter (1966). Multiple forms of fructose diphosphate aldolase in mammalian tissues. *Proc. Nat. Acad. Sci.*, 56:1275–1282.

192. Periman, Ph. (1970). IgG synthesis in hybrid cells from an antibody-producing mouse myeloma and an L cell substrain. *Nature*, 228:1086–1087.

193. Peters, T. Jr. (1970). Serum albumin, *Adv. Clin. Chemistry*, 13:37–111.

194. Peterson, J. and M. C. Weiss (1971). Expression of differentiated functions in hepatoma cell hybrids. Induction of mouse albumin synthesis in rat hepatoma-mouse fibroblast hybrids. *Proc. Nat. Acad. Sci., USA*, 69:51–75.

195. —— and M. C. Weiss (1971). Expression of differentiated functions in hepatoma cell hybrids. VI. Quantitative studies of albumin synthesis (in preparation).

196. Pfeiffer, S. E., H. R. Herschman, J. Lightbody, and G. Sato (1970). Synthesis by a clonal line of rat glial cells of a protein unique to the nervous system. *J. Cell. Physiology*, 75:329–340.

197. Pfeiffer, S. E., H. R. Herschman, J. E. Lightbody, G. Sato, and L. Levine (1971). Modification of cell surface antigenicity as a function of culture conditions. *J. Cell. Physiology*, 78:145–152.

198. Pierce, G. B. (1967). Teratocarcinoma: model for a developmental concept of cancer. In *Current Topics in Develop-*

mental Biology (A. A. Moscona and A. Monroy, eds.) Vol. 2, pp. 223–246. Academic Press, New York.

199. Pitot, H. C., C. Peraint, P. A. Morse, Jr., and V. R. Potter (1964). Hepatomas in tissue culture compared with adapting liver *in vivo*. In *Metabolic Control Mechanisms in Animal Cells*, pp. 229-246. National Cancer Institute Monograph 13.

200. Pollack, R., H. Green, and G. J. Todaro (1968). Growth control in cultured cells: selection of sublines with increased sensitivity to contact inhibition and decreased tumor-producing ability. *Proc. Nat. Acad. Sci., USA,* 60:126–133.

201. ———, S. Wolman, and A. Vogel (1970). Reversion of virus transformed cell lines: hyperploidy accompanies retention of viral genes. *Nature,* 228:938–970.

202. Pontecorvo, G. (1961). Genetic analysis via somatic cells. *The Scientific Basis of Medicine Ann. Reviews,* 13–20.

203. ——— (1971). Induction of directional chromosome elimination in somatic cell hybrids. *Nature,* 230:367–369.

204. Poole, A. R., J. I. Howell, and J. A. Lucy (1970). Lysolecithin and cell fusion. *Nature,* 227:810–814.

205. Puck, T. T. and F. Kao (1967). Genetics of somatic mammalian cells. V. Treatment with 5-bromodeoxyuridine and visible light for isolation of nutritionally deficient mutants. *Proc. Nat. Acad. Sci., USA,* 58:1227–1234.

206. Rabinowitz, Z. and L. Sachs (1968). Reversion of properties in cells transformed by polyoma virus. *Nature,* 220:1203–1206.

207. Rao, P. N. and R. T. Johnson (1970). Mammalian cell fusion. I. Studies on the regulation of DNA synthesis and mitosis. *Nature,* 225:159–164.

208. Ricciuti, F. and F. H. Ruddle (1971). Biochemical and cytological evidence for triple hybrid cell line formed from fusion of three different cells. *Science,* 172:470–472.

209. Richler, C. and D. Yaffé (1970). The *in vitro* cultivation and differentiation capacities of myogenic cell lines. *Devel. Biol.,* 23:1–22.

210. Ringertz, N. R. and L. Bolund (1969). "Activation" of hen erythrocyte deoxyribonucleoprotein. *Exp. Cell Res.,* 55: 205–214.

211. Ruddle, F. H., V. M. Chapman, T. R. Chen, and R. J. Klebe (1970). Linkage between human lactate dehydrogenase A and B and peptidase B. *Nature*, 227:251–257.

212. ———, T. Chen, T. B. Shows, and S. Silagi (1970). Interstrain somatic cell hybrids in the mouse. Chromosome and enzyme analyses. *Exp. Cell Res.*, 60:139–147.

213. ——— and L. Harrington (1967). Tissue specific esterase isozymes of the mouse (*Mus musculus*). *J. Exp. Zool.* 166: 51–64.

214. Sanford, K. K., G. D. Likely, and W. R. Earle (1954). The development of variations in transplantability and morphology within a clone of mouse fibroblasts transformed to sarcoma producing cells *in vitro. J. Nat. Cancer Instit.* 23: 1035–1059.

215. Santachiara, A. S., M. Nabholz, V. Miggiano, A. J. Darlington, and W. Bodmer (1970). Genetic analysis with man-mouse somatic cell hybrids. *Nature*, 227: 248–251.

216. Sato, G. H. and V. Buonassisi (1964). Hormone secreting cultures of endocrine tumor origin. In *Metabolic Control Mechanisms in Animal Cells*, pp. 81-89. Nat. Cancer Inst. Monograph No. 13.

217. Scaletta, L. and B. Ephrussi (1965). Hybridization of normal and neoplastic cells *in vitro. Nature*, 207:1169–1171.

218. ———, N. B. Rushforth, and B. Ephrussi (1967). Isolation and properties of hybrids between somatic mouse and Chinese hamster cells. *Genetics*, 57:107–124.

219. Schneeberger, E. E. and H. Harris (1966). An ultrastructural study of interspecific cell fusion induced by inactivated Sendaï virus. *J. Cell Sci.*, 1:401–406.

220. Schneider, J. A. and M. C. Weiss (1971). Expression of differentiated functions in hepatoma cell hybrids. I. Tyrosine aminotransferase in hepatoma-fibroblast hybrids. *Proc. Nat. Acad. Sci., USA,* 68:127-131.

221. Schubert, D. and F. Jacob (1970). 5-bromodeoxyuridine-induced differentiation of a neuroblastoma. *Proc. Nat. Acad. Sci., USA,* 67:247–254.

222. Schwartz, A. G., P. R. Cook, and H. Harris (1971). Correction of a genetic defect in a mammalian cell. *Nature New Biol.*, 230:5–8.

164

223. Sherwood, L. M., J. T. Potts, Jr., R. A. Melick, and G. D. Auerbach (1966). Parathyroid hormone production by bronchogenic carcinomas. (Abstract.) *Program of the 48th Meeting of the Endocrine Soc.*, p. 29. J. B. Lippincott Co., Philadelphia, Pa.

224. Sidebottom, E. and H. Harris (1969). The role of the nucleolus in the transfer of RNA from nucleus to cytoplasm. *J. Cell Sci.*, 5:351–364.

225. Silagi, S. (1967). Hybridization of a malignant melanoma cell line with L cells *in vitro*. *Cancer Res.*, 27:1953–1960.

226. —— and S. A. Bruce (1970). Suppression of malignancy and differentiation in melanotic melanoma cells. *Proc. Nat. Acad. Sci., USA,* 66:72–78.

227. ——, G. Darlington, and S. A. Bruce (1969). Hybridization of two biochemically marked human cell lines. *Proc. Nat. Acad. Sci., USA,* 62:1085–1092.

228. Silvestri, L. G., ed. (1971). *The Biology of oncogenic viruses.* North Holland Publishing Co., Amsterdam.

229. Siniscalco, M., H. P. Klinger, H. Eagle, H. Koprowski, W. Y. Fujimoto, and J. E. Seegmiller (1969). Evidence for intergenic complementation in hybrid cells derived from two human diploid strains each carrying an X-linked mutation. *Proc. Nat. Acad. Sci., USA,* 62:793–799.

230. ——, B. B. Knowles, and Z. Steplewski (1969). Hybridization of human diploid strains carrying X-linked mutants and its potential for studies of somatic cell genetics. In *Heterospecific Genome Interaction,* The Wistar Institute Symp. Monogr. No. 9 (V. Defendi, ed.), pp. 117–133. The Wistar Institute Press, Philadelphia.

231. —— (1970). Somatic cell hybrids as tools for genetic studies in man. In *Control Mechanisms in the Expression of Cellular Phenotypes* (H. A. Padykula, ed.), Academic Press, New York.

232. Sobel, J. S., A. M. Albrecht, H. Rierm, and J. L. Biedler (1971). Hybridization of Actinomycin D- and Amethopterin-resistant Chinese hamster cells *in vitro*. *Cancer Res.*, 31:297–307.

233. Sonneborn, T. M. (1964). The differentiation of cells. *Proc. Nat. Acad. Sci., USA,* 51:915–929.

234. Sonneborn, T. M. (1970). Gene action in development. *Proc. Roy. Soc. Lond. B*, 176:347–366.

235. Sonnenschein, C. (1969). Discussion. In *Heterospecific Genome Interaction*, The Wistar Institute Symp. No. 9 (V. Defendi, ed.), p. 111. The Wistar Institute Press, Philadelphia.

236. ———, D. Robertz, and G. Yerganian (1969). Karyotypic and enzymatic characteristics of a somatic cell line orginating from dwarf hamsters. *Genetics*, 62:379–392.

237. ———, A. Tashjian, and U. Richardson (1968). Somatic cell hybridization: mouse-rat hybrid cell line involving a growth hormone producing parent. *Genetics* (Abstract) 60:227.

238. Sorieul, S. and B. Ephrussi (1961). Karyological demonstration of hybridization of mammalian cells *in vitro*. *Nature*, 190:653–654.

239. Spencer, R. A., T. S. Hauschka, D. B. Amos, and B. Ephrussi (1964). Co-dominance of isoantigens in somatic cells grown *in vitro*. *J. Nat. Cancer Instit.*, 33:893–903.

240. Spiegelman, S. (1957). Nucleic acids and the synthesis of proteins. In *The Chemical Basis of Heredity* (William D. McElroy and Bentley Glass, eds.), pp. 232–267. The Johns Hopkins Press, Baltimore.

241. Stanners, C. P., G. L. Eliceiri, and H. Green (1971). Two types of ribosome in mouse-hamster hybrid cells. *Nature New Biol.*, 230:52–54.

242. Stent, G. S. (1969). *The Coming of the Golden Age: a View of the End of Progress*. Natural History Press, Gordon City, N.Y.

243. Szybalski, W., E. H. Szybalska, and G. Regnie (1962). Genetic studies with human cell lines. *National Cancer Inst. Monogr.* No. 7:75–89.

244. Tartar, V. (1961). *The Biology of Stentor*. Pergamon Press, New York.

245. Tashjian, A., Y. Yasumura, L. Levine, G. Sato, and M. Parker (1968). Establishment of clonal strains of rat pituitary tumor cells that secrete growth hormone. *Endocrinology*, 82:342–352.

246. Thompson, E. B. and T. D. Gelehrter (1971). Expression of tyrosine aminotransferase activity in somatic cell hetero-karyons: evidence for negative control of enzyme expression. *Proc. Nat. Acad. Sci., USA*, 68:2589–93.

247. Tomkins, G. M., T. D. Gelehrter, D. Granner, D. Martin, Jr., H. Samuels, and E. Brad Thompson (1969). Control of specific gene expression in animal cells. *Science*, 166:1474–1480.

248. Tsanev, R. and Bl. Sendov (1971). Possible molecular mechanism for cell differentiation in multicellular organisms. *J. Theor. Biol.*, 30:337–393.

249. Wasserman, E. and L. Levine (1961). Quantitative micro-complement fixation and its use in the study of antigenic structure by specific antigen-antibody inhibition. *J. Immunology*, 87:290–295.

250. Watkins, J. F. and R. Dulbecco (1967). Production of SV40 virus in heterokaryons of transformed and susceptible cells. *Proc. Nat. Acad. Sci., USA*, 58:1396–1403.

251. ——— and D. M. Grace (1967). Studies on the surface antigens of interspecific mammalian cell heterokaryons. *J. Cell Sci.*, 2:193–204.

252. Weiss, L. (1967). *The Cell Periphery, Metastasis and Other Contact Phenomena*. North Holland Publ. Co., Amsterdam.

253. Weiss, M. C. (1970). Further studies on loss of T-antigen from somatic hybrids between mouse cells and SV40-trans-formed human cells, *Proc. Nat. Acad. Sci., USA*, 66:79–86.

254. ——— (1970). Properties of somatic hybrid cell lines be-tween mouse cells and SV40-transformed human cells. In *Genetic Concepts and Neoplasia* (The University of Texas M. D. Anderson Hospital and Tumor Institute at Houston, 23rd Annual Symposium on Fundamental Cancer Research, 1969), pp. 456–476. Williams and Wilkins, Baltimore.

255. ——— and M. Chaplain (1971). Expression of differen-tiated functions in hepatoma cell hybrids. III. Re-expression of tyrosine aminotransferase inducibility, *Proc. Nat. Acad. Sci., USA*, 68:3026–30.

256. ——— and B. Ephrussi (1966). Studies of interspecific (rat × mouse) somatic hybrids. I. Isolation, growth, and evolution of the karyotype. *Genetics*, 54:1095–1108.

257. Weiss, M. C. and B. Ephrussi (1966). Studies of interspecific (rat × mouse) somatic hybrids. II. Lactate dehydrogenase and β-glucuronidase. *Genetics*, 54:1111–1122.

258. ———, B. Ephrussi, and L. J. Scaletta (1968). Loss of T-antigen from somatic hybrids between mouse cells and SV40-transformed human cells. *Proc. Nat. Acad. Sci., USA*, 59:1132–1135.

259. ——— and H. Green (1967). Human-mouse hybrid cell lines containing partial complements of human chromosomes and functioning human genes. *Proc. Nat. Acad. Sci., USA*, 58:1104–1111.

260. ———, G. J. Todaro, and H. Green (1968). Properties of a hybrid between lines sensitive and insensitive to contact inhibition of cell division. *J. Cell. Physiology*, 71:105–107.

261. Weiss, P. (1967). Discussion. In *Cell Differentiation*, Ciba Foundation Symp. (A.V.S. De Reuck and J. Knight, eds.), p. 13. J. and A. Churchill, Ltd., London.

262. Wiener, F., G. Klein, and H. Harris (1971). The analysis of malignancy by cell fusion. III. Hybrids between diploid fibroblasts and other tumor cells. *J. Cell Sci.*, 8:681–692.

263. Wildemuth, H. (1970). Determination and transdetermination in cells of the fruitfly. *Sci. Progress, Oxford*, 58:329–358.

264. Wolstenholme, G.E.W. and J. Knight, eds. (1971). *Growth Control in Cell Cultures* (Ciba Symposium), Churchill Livingstone, Edinburgh and London.

265. Yaffé, D. (1968). Retention of differentiation potentialities during prolonged cultivation of myogenic cells. *Proc. Nat. Acad. Sci., USA*, 61:477–483.

266. Yerganian, G. and M. Nell (1966). Hybridization of dwarf hamster cells by UV-inactivated Sendaï virus. *Proc. Nat. Acad. Sci., USA*, 55:1066–1073.

267. Yoshida, M. C. and B. Ephrussi (1967). Isolation and karyological characteristics of seven hybrids between somatic mouse cells *in vitro*. *J. Cell. Physiology*, 69:33–44.

Author Index

Numbers indicate pages where an author's name or a reference to his work is to be found.

171

Subject Index

Acetabularia, 115

albumin, 84; induction of synthesis by fibroblast genome, 138-139; partial extinction of in hepatoma hybrids, 86-87, 139, 141; secretion by hepatoma cells, 84

aldolase B, extinction of in hepatoma hybrids, 85-86, 88; B, re-expression of, 91-92; tissue distribution of isozymes of, 83-84

aminopterin, 13

antibody, production by hybrids, 139-141

antigens, in hybrid cells, 9; inheritance of virus induced, 99

8-azaguanine, resistance to, 11

bacteriophages, 115

β-glucuronidase in hybrids, 9, 27

bipartite inheritance, 115

bromodeoxyuridine, cell surface changes caused by, 108; effect on differentiation and malignancy, 108; mechanism of action on differentiation, 108; resistance to, 12

cell, cortex, 114-116; culture, techniques, 5; membrane, 54-55; membrane, specificity of, 144-145; strains, definition and properties, 118

chromatin, condensation of, and nuclear re-activation, 61-63

chromosome (s), as "markers" of parental cells, 6-7; directed loss, 127; human, identification by cytochemical techniques, 128-130; mechanism of loss, 121-125; non-randomness of loss, 128; preferential loss, 126; reversal of preferential loss, 124; pulverization, 46-47, 120, 122; translocation T-6 as "marker," 11

collagen, production by hybrid cells, 141

complementation, 13, 134

contact inhibition, 11, 89-90, 145; inheritance of, 99

cortex of eggs, role in development, 145

cortical inheritance, 145

dedifferentiation, 51

determination, 52-53; as a change of intrinsic cell differences *vs.* group stability, 111; role of cytoplasm in, 70; stability of, 53

development, guidance by genes *vs.* cell cortex, 114-116; primary causes of, 113-116

differentiation, 52; and determination, mechanisms of, 131-133; heritability of, 52, 54-55; maintenance of, *in vitro*, 50-52; mechanism of, 110-112; of diploid cells *in vitro*, 51-52; of malignant cells *in vitro*, 50-51

DNA synthesis, suppression of, in HeLa-Ehrlich heterokaryons, 22

dogma, 113; unwritten, 114-115

dopa oxidase, in melanoma cells, 73; extinction of, in hybrids, 73-77; inhibitor of, 76

dormant nuclei, mechanism of reactivation, 62

effective mating rate, 14

epigenetic, changes, seat of, 95-96; states, mechanism of, 53-56

epigenotype, 53; and regulation of bacterial type in differentiation, 112; as regulatory state, 54; basis of, 54-56; changes of, 138; nuclear DNA as seat of, 55

erythrocyte heterokaryons, 57-70; production of hemoglobin by, 67-70

erythrocyte nuclei, activation in heterokaryons, 59-67; production of antigens, 64; production of HGPRT, 65-67

erythrocytes, lysis by Sendaï virus, 59

esterase ES-2, production by kidney adenocarcinoma, 82; extinction of in RAG × fibroblast hybrids, 94-95; re-expression correlated with loss of chromosomes, 94-95

extinction of luxury functions, as caused by diffusible regulator

173